neurophysiol

neurophysical

SOCIETY FOR NEUROSCIENCE SYMPOSIA
Volume III

Society for Neuroscience

Officers, 1977

President	Floyd E. Bloom
President-Elect	W. Maxwell Cowan
Secretary	David H. Cohen
Treasurer	Jennifer S. Buchwald

1977 Program Committee

Robert W. Doty, *Chairman*
Jesus Alanis
Jeffery L. Barker
Reginald G. Bickford
William H. Calvin
Sven O. E. Ebbesson
Ann M. Graybiel
J. Allan Hobson

Arthur J. Hudson
Lawrence Kruger
John I. Lacey
Frederick A. Miles
Richard J. Wurtman
Floyd E. Bloom, *ex officio*
David H. Cohen, *ex officio*

1977 Communication Committee

W. Maxwell Cowan, *Chairman*
Richard P. Bunge

James A. Ferrendelli
Walle J. H. Nauta

SOCIETY FOR NEUROSCIENCE SYMPOSIA
Volume III

ASPECTS OF BEHAVIORAL NEUROBIOLOGY

James A. Ferrendelli, M.D., Editor
Department of Pharmacology and Neurology
Washington University School of Medicine

Assistant Editor: Gerry Gurvitch
Society for Neuroscience

Published by the Society for Neuroscience
Bethesda, Maryland

Library of Congress Catalogue Card Number 75-27110

International Standard Book Number 0-916110-06-0

Papers presented at two symposia during the
Seventh Annual Meeting of the
Society for Neuroscience, held in Anaheim, California, 1977.

PREFACE

This is the third volume in the series of *Society for Neuroscience Symposia*. As in the past, only a few of the several excellent symposia presented at the Society's Annual Meeting are selected for publication. Our aim is to have a central theme in each volume and to include material which more or less adheres to this theme. However, because of the great diversity in neurosciences, and for many other reasons, this is not always possible. Thus, the present volume, entitled *Aspects of Behavioral Neurobiology*, contains papers from two symposia: "Neural Correlates of Behavioral State" and "Neurochemistry of Olfactory Circuits." These are subjects not covered in previous volumes and, I am pleased to say, are very exciting areas which should be of interest to most neuroscientists.

The symposium "Neural Correlates of Behavioral State" was organized and chaired by J. Allan Hobson. This section begins with a chapter by Dr. Hobson answering the question "What Is a Behavioral State?" His comments are useful and very timely and set the stage for subsequent chapters. The next four chapters deal with physiologic and chemical correlates of behavior in both simple and complex systems. Two consider the well-defined behaviors of gill-withdrawal and feeding in *Aplysia*. The other two discuss two aspects of sleep in the cat— "State-Dependent Reversal of a Brainstem Reflex" and "Control of Sleep-Waking State Alteration."

The second symposium, "Neurochemistry of Olfactory Circuits," was organized and chaired by Gordon M. Shepherd. He outlines its goals and content very well in his introduction, so I will not repeat that information here. The symposium represents one of the most comprehensive and sophisticated considerations of the neurochemistry and neuropharmacology of olfaction and olfactory circuits available. There is much recent data presented in the six chapters in this section, including specific concepts about the olfactory system and demonstrations of elegant techniques which are applicable in many other areas of the nervous system.

I am certain that the present volume will be as successful as the previous volumes in this series. My contribution and that of the Communication Committee is relatively small, and many others deserve recognition for their efforts. First, our thanks go to the organizers and participants of the symposia. I greatly appreciate their participation in the meeting and careful preparation of manuscripts. Of course we thank Mrs. Marjorie Wilson, Executive Secretary of the Society, and Gerry Gurvitch, assistant editor, who are the ones really responsible for the development and production of each volume. Finally, we wish to thank Mrs. Hanne Caraher, Ms. Dorothy A. Kinscherf, and Mrs. Doris Stevenson for secretarial assistance and help in the editing of this volume.

James A. Ferrendelli
St. Louis, March 1978

CONTENTS

PARTICIPANTS

Claire Advokat
Division of Neurobiology and
  Behavior
Columbia University
College of Physicians and
  Surgeons
New York, New York 10032

Richard D. Broadwell
Laboratory of Neuropathology
  and Neuroanatomical Sciences
National Institute of Neurological
  and Communicative Disorders
  and Stroke, NIH
Bethesda, Maryland 20014

Michael H. Chase
Department of Physiology
Center for Health Sciences
University of California
Los Angeles, California 90024

Robert D. Hawkins
Division of Neurobiology and
  Behavior
Columbia University
College of Physicians and
  Surgeons
New York, New York 10032

J. Allan Hobson
Laboratory of Neurophysiology
Department of Psychiatry
Harvard Medical School
Boston, Massachusetts 02115

Stephen Hunt
Department of Psychiatry and
  Behavioral Science
State University of New York at
  Stony Brook
Stony Brook, New York 11794

Irving Kupfermann
Division of Neurobiology and
  Behavior
Columbia University
College of Physicians and
  Surgeons
New York, New York 10032

Frank L. Margolis
Department of Physiological
  Chemistry and Pharmacology
Roche Institute of Molecular
  Biology
Nutley, New Jersey 07110

Robert W. McCarley
Laboratory of Neurophysiology
Department of Psychiatry
Harvard Medical School
Boston, Massachusetts 02115

Donald W. Pfaff
The Rockefeller University
New York, New York 10021

Charles E. Ribak

Division of Neurosciences
City of Hope National Medical
  Center
Duarte, California 91010

Jakob Schmidt

Department of Biochemistry
State University of New York
  at Stony Brook
Stony Brook, New York 11794

Gordon M. Shepherd

Department of Physiology
Yale University School of Medicine
New Haven, Connecticut 06510

Klaudiusz R. Weiss

Division of Neurobiology and
  Behavior
Columbia University
College of Physicians and
  Surgeons
New York, New York 10032

NEURAL CORRELATES OF
BEHAVIORAL STATE

WHAT IS A BEHAVIORAL STATE?

J. Allan Hobson

Harvard Medical School, Boston, Massachusetts

THE STRATEGY OF SIMPLICITY AND THE STATE CONCEPT

Two symmetrical strategies shape behavioral neurobiology today. One is the study of simple nervous systems. The other is the study of simple behaviors. Together they comprise what I will call the "Strategy of Simplicity." When combined with the precise methods of cellular neurophysiology, as in the work of Kandel (1976) and others on the sea snail *Aplysia,* they have proven extremely fruitful.

In animals with more complex nervous systems and more complex behaviors—such as the vertebrates—it is likely to be true that the choice of relatively simple behaviors, controlled by relatively discrete parts of the nervous system, will also be propitious. Thus it is this same Strategy of Simplicity that has attracted many mammalian behavioral neurobiologists to the study of sleep. As Evarts (1967) has argued, we must try to understand sleep before we can hope to understand the more complex behaviors of the awake animal. Guided by this rationale, Evarts (1967) pioneered the application of the single-cell approach to the study of sleep in the house cat *Felix* and developed quantitative measures to characterize state-dependent changes in neuronal activity. Simultaneously, Jouvet (1962) demonstrated that the neuronal machinery for generating the feline sleep cycle was localized in the pontine brainstem. Now the brainstem—which Ramón y Cajal (1937) called a "wearisome labyrinth"—is certainly not simple when compared to the abdominal ganglion of *Aplysia,* but Jouvet (1962) showed that it could at least be studied in the absence of the massive and complex forebrain.

The stage was thus set for an attempt to analyze the cellular mechanism of sleep cycle control by applying both components of the Strategy of Simplicity (Hobson, 1974; Steriade and Hobson, 1976). But, here and elsewhere in behavioral neurobiology, the productive application of the Strategy of Simplicity depends upon the parallel development of methods and concepts in not only the neurobiological but also the behavioral domain. The paradigms of experimental psychology—especially those comprising learning theory—have been too limited to allow full exploitation of the rich armamentarium of neurobiological techniques, especially those which analyzed events at the cellular level. Also conspicuously neglected by classical experimental psychology have been those behaviors which are innate and spontaneous. This is unfortunate since among the innate, spontaneous behaviors are the states of arousal and sleep, which strongly invite study using the Strategy of Simplicity.

ETHOLOGY AND THE STATE CONCEPT

Even the birth and development of ethology—the naturalistic study of behavior—which has broadened our view of behavior in ways quite useful to behavioral neurobiology, has not yet provided data or paradigms for studying the behavioral states. As yet, there is no ethology of "state." It is probable that ethologists have eschewed this potentially fertile ground for two related historical reasons: One is the unfortunate history of the "instinct" concept and the ghost of vitalism that hovered around it (Lashley, 1938); the other is the association of sleep research with the teleogically contaminated concept of "motivational state" (Tolman, 1932). Both the instinct and motivational concepts relegated state to unobservable and thus unmeasureable phenomena—the so-called intervening variables. Hebb's (1955) concept of "drive strength" helped by allowing quantification of behavior, but it did not lead to analytic paradigms of use to neurobiologists.

In addition to these negative factors, the ethologists had good operational reasons for concentrating their attention on those behavioral phenomena that were stimulus-bound and ended in organized motor acts: such behaviors could be readily recognized, quantified, and experimentally manipulated. The purview of ethology is thus today limited to one set of state-dependent behaviors, those occurring in waking. Neglected is fully one-third of mammalian behavior and, more important, a way of dealing with state variables as they influence all behavior.

In Hinde's classic and rigorous textbook *Animal Behavior* (1966), there is not a single reference to sleep, and although "state" is used repeatedly throughout the book as if it meant "an internal variable that influences the probability of occurrence of behavior," the term is never formally defined and behavioral paradigms for its study are not discussed. Yet it is clear to everyone that sleep is a behavior as real and compelling as sexuality, nest building, and feeding, behaviors that fit more easily into the mold of the ethological version of the stimulus-response paradigm: an external stimulus is a "releaser" which triggers a response from "innate releasing mechanisms." Moruzzi has argued convincingly that sleep can be considered as an instinctive behavior, even as classically defined (1969). Moreover, the REM phase of mammalian sleep is as clear an example of a fixed action pattern as can be found in mammals. REM sleep has the full complement of Lorenzian (1937) criteria for a fixed action pattern: central program, stereotypy, and genetic determinism (see also McCarley, 1978).

These accommodations to the existing framework, while progressive, are still inadequate. In order to deal with such automatic and enduring behaviors as sleep, I submit that we need a wholly new and different conceptual approach, one that is capable of defining, measuring, and analyzing "state." No theory of behavior can be complete without it, since, as is widely acknowledged, state determines the way in which the reflexes of the animal are organized to produce all of its behavior. Such a global phenomenon might seem, at first glance, to violate the Strategy of Simplicity. On closer inspection the automaticity, stereotypy, and localized control of states suggest that their analysis is not only necessary but feasible using the Strategy of Simplicity and the potent cellular techniques of neurobiology.

THE STIMULUS-RESPONSE PARADIGM AND THE STATE CONCEPT

We have mentioned that applying the Strategy of Simplicity to behavior evolved, first and foremost, the stimulus-response paradigm of experimental psychology, which matched the powerful Sherringtonian reflex concept in neurobiology (1906). The Sherrington concept of the integrative action of the nervous system and the paradigms of learning theory developed *pari passu* and with mutual enrichment. In its simplest form, a stimulus (S) is presented to the organism, it is operated upon by the central nervous system (CNS), and a response (R) is generated. Thus, the reflex paradigm:

Reflex: $$S \rightarrow CNS \rightarrow R$$

In what has been considered by the overwhelming majority of investigators to be the next logical step, the stimulus is varied in strength, in quality, in number, in order, and in time as an indirect means of ascertaining the operation performed on the stimulus by the CNS. Thus the variants of the reflex paradigm include:

Intensity:	$2S \to CNS \to 2R$
Quality:	$S_1 \ldots S_2 \to CNS \to R_{1,2}$
Number:	$S_1 \ldots S_1 \to CNS \to R_2$
Order:	$S_2 \ldots S_1 \to CNS \to R_{2,1}$
Interval:	$S_2 \ldots \ldots S_1 \to CNS \to R_1$

Notice that the interpretation of even these relatively simple experiments assumes that the CNS is the same CNS in all cases. Efforts routinely taken to hold the "state" of the system constant include excision and anesthesia, two suppressive treatments with their own dramatically dynamic consequences.

But an assumption of the learning theorists who take the first step in extending the Sherringtonian concept to behavior is that the CNS *changes* from trial to trial as a consequence of the trials. Thus, in the habituation paradigm:

Habituation:	$S_1 \to CNS_1 \to R_1$
	$S_1 \to CNS_2 \to R_2 \ (R_2 < R_1)$

In other words, the plasticity of the system is a widely accepted corollary of the stimulus-response paradigm, and the assumption of constancy is challenged even within the restricted conditions of acute experiments.

In addition to these experimentally induced changes, the state of the system is constantly changing *spontaneously*. To demonstrate this, one has only to apply the alluringly simple *state* paradigm: keep S constant and observe the changes in R as function of time:

State:	$S_1 \ (T_0) \to CNS_1 \to R_1$
	$S_2 \ (T_1) \to CNS_2 \to R_2$
	$S_3 \ (T_2) \to CNS_3 \to R_3$

Or better yet, eliminate S and the stimulus-response paradigm altogether:

State:	$T_0 \ CNS_1$
	$T_1 \ CNS_2$
	$T_2 \ CNS_3$

Perhaps it is because this state paradigm is so simple and obvious that it has been overlooked. Yet it provides a valuable approach to the documentation and interpretation of the intrinsic dynamism of the nervous system. The state of the system, as it turns out, is highly variable but also highly predictable. State is thus a basic and critical variable of enormous interest in its own right and of clear significance in interpreting the results of any stimulus-response experiment. The interaction of state and plasticity is so commonplace in our experience that it is truly remarkable that it has, as yet, no place in our experiments. State variables are always at least as significant as stimulus variables and sometimes even more so. The goal of this symposium was to review current work on neuronal correlates of behavioral state with the hope of formulating definitions, paradigms, and interpretive logic to aid the development of this important field.

BEHAVIORAL NEUROBIOLOGY AND THE STATE CONCEPT

Let us now attempt to go beyond the stimulus-response paradigm that has so effectively guided research in neurobiology for almost a century. Careful development of the state concept is crucial to mammalian behavioral neurobiology at the lowest suprasegmental levels—that of the brainstem, for example. For those who work on submammalian systems, substitution of in-series circuits with short time-constant elements for "spinal cord level" and substitution of in-parallel circuits with long time-constant elements for "brainstem" will generalize the problem and the approach developed here to most known cases.

Let us quickly note that the reflex concept and the state concept are not mutually exclusive. Indeed they are complementary. But they are not the same, and it is critical for the experimentalist to recognize the differences because he needs a special attitude and approach to each problem. In the extreme, the state concept involves the notion of "spontaneity," or relative independence of input, of controlling neural systems. Of course no neural system is independent of input, but state control systems are responsive to input in such a nonspecific sense when compared to the specificity of reflex activity as to make an almost qualitative difference.

Indeed, the scientist wishing to study behavioral states and their neural correlates may find himself in the awkward position of performing "non-experiments"—using Claude Bernard's definition of an experiment as the norm (1865). Our training, entirely consistent with the stimulus-response paradigm, is to establish constant conditions, to stimulate, to record responses under control conditions, to change some variable, to record responses under experimental conditions, and so forth. This approach may be frustrated by the reality of constant state change and will not work at all in its scientific investigation.

This is because it is an *intrinsic change in the system* that is the object of investigation. That is, a state change, which constitutes a variable to be eliminated or controlled by the S-R scientists, must at first be left unhindered by the state physiologist. In other words, the experimenter, to investigate state, must keep himself constant and let the observed system run free if he wishes to obtain even the simplest descriptive information. This apparently trivial, transparent point actually constitutes a major obstacle to the classically trained investigator who, following experimental traditions, wants to be the state-regulator himself: it is his actions which produce the most significant state changes in the system under study. Thus, when the experimenter gives anesthesia or performs a decerebration, he may eliminate the very object that the state neurobiologist wishes to study.

Let us now consider some of the differences between acute and chronic experiments. Acute experiments are not only too disruptive to state control mechanisms, but also too short to allow state evolution to be visualized. It used to be that chronic experimentation involved doing things (experiments) to animals that were unphysiologic, as many acute experiments still are (i.e., electrically stimulating or lesioning the brain). This is no longer true. Chronic recording studies may now be considered as among the most physiologically pure investigations, even if they are not experiments in Claude Bernard's sense of the word.

If one is to cut himself off from the secure moorings of the stimulus-response paradigm, he needs to give some thought to the uncharted sea he means to explore. With adequate concepts and methods he may then come back to the safe harbor with a new view of the physiological world. He may even then be able to incorporate the powerful S-R paradigm into his second, more exploitative voyage into new waters. The first great need is for definitions. What is a state? How will we recognize a state when we see it? How can we measure a state when we see it? What meaning will we attach to our data? A synthesis of the responses to these questions given by the contributors to this symposium follows.

WHAT IS A STATE?

State, at the most general level, is any condition of a system. More precisely, and following Ashby (1960), we all agree that "The state of a system at a given instant is the set of numerical values which its variables have at that instant.''

A completely specified system state would include all possible measurements on the system and predict the effect of all inputs to the system in terms of output and changes in state. A complete description of state would also take account of any intrinsic unstable properties of the system leading to state change as well as the means by which such changes are compensated or regulated. Since a complete specification (as in finite automata theory) can never be achieved for biological systems, most investigators use some arbitrary set of variables which can be objectively measured. In the initial stages of investigation, this is acceptable if the degree of resolution matches the question asked of the system, but, as Ashby points out, any set of variables selected by an observer from the infinite set available may create a picture quite different from that of the ''real machine.'' An even more critical problem is that the variables chosen to characterize state must be appropriate to the level of analysis intended. Thus a *behavioral* state is defined behaviorally, not physiologically, while a *physiological* state is defined physiologically and not behaviorally. If this criterion is not carefully observed, circularity is introduced into the analysis of the data and we risk viewing states as causing themselves.

WHAT BEHAVIORAL STATES CAN BE STUDIED AND HOW ARE THEY OBJECTIFIED AND MEASURED?

In most animals the behavioral states of activity and rest are distinguished by a set of variables as in Table 1. In mammalian vertebrates we refer to the active state as waking and distinguish two major substates of sleep: NREM and REM. In *Aplysia* the criteria for ''awake'' apply to ''food arousal'' while the criteria for ''asleep'' apply to ''nonfood arousal.'' There is no clear equivalent of REM in *Aplysia*.

In mammals these behavioral conditions have a set of physiological correlates which are so constantly concomitant with the behavior that

TABLE 1. *Behavioral "states"*

State criterion	Activity (awake)	Rest (asleep)	
		NREM sleep	REM sleep
Posture	Erect	Recumbent	Flaccid
Mobility	High, flexible	Low	Limited and stereotyped
Eyes	Open and moving	Closed	Closed and moving
Responsiveness	High	Low	Lower

they have become interchangeable. But it should be made clear that when we make the observations listed in Table 2 we are measuring physiological, not behavioral, states.

Most investigators now regard the one-to-one correspondence of these correlated behavioral states and physiological states as an identity. But, in so doing, it must be kept clear that nothing has been explained. The only advantage gained is ease of objective identification and quantification.

The previously stated caveats regarding the degree of resolution needed and the level at which the state concept is applied come more strongly into focus when we choose to examine state-dependent behaviors such as feeding, which we would all agree only occurs in waking (or "food arousal," since the term "waking" may not be appropriate in invertebrates). Whether feeding itself should be regarded as a "state" is questionable; it would seem to be more properly viewed as a consummatory response to an inferred motivational state called "hunger" (which we cannot measure) or the appetitive behavioral state of "food seeking" (which can be quantified). What is the essential difference between behavioral and motivational states? The former have no goal or termination point outside the organism, and they appear to be regulated autonomously. The latter involve external objects which can terminate the state. Thus in *Aplysia* the state of food seeking can be identified and measured as the magnitude and latency of general locomotor responses and specific biting responses to food presentation. In this sense the food-seeking (or food-deprived) state is analogous to the sleep-deprived state in mammals whose parameters of measurement (latency to sleep onset, EEG amplitude and frequency, etc.) are formally identical to those used by the invertebrate physiologists. This analogy raises the possibility, first taken seriously by Moruzzi (1969), that sleep, like feeding, is more properly viewed as a consummatory response rather than as a state; in this view, "being sleepy" is an appetitive phase analogous to food seeking, but the consummatory response, as noted above, does not involve an external event or object.

TABLE 2. *Physiologic "states"*

	Awake	NREM sleep	REM sleep
EMG	+ +	+	−
EEG	+	−	+
EOG	+	−	+ +

It is interesting to note that the snail makes its "hungry state" known spontaneously just as the mammal would. When *Aplysia* is "hungry" (food-deprived), it increases locomotor activity and shows bursts of spontaneous head waving; some animals, long starved, even show spontaneous biting movements (see Weiss and Kupfermann, 1978). Thus, only the degree to which states are stimulus-bound appears to distinguish the invertebrates from the vertebrates. On this basis some might even question the use of the term "state" to describe the food-seeking condition of *Aplysia*. Convincing justification for the use of the term seems to lie in the fact that, once aroused, the clearly defined, measurable, food-seeking state persists for up to 30 minutes. In addition, we know from Strumwasser's work that arousal and dormancy do alternate "spontaneously" in the snail, and we are probably correct in assuming that "food seeking" is a substate of arousal under these conditions (1971).

TEMPORAL ASPECTS OF STATES

The quality of durability seems to be central to the application of the state concept to the study of behavior. Duration is always implied when we say the animal is in one or another state. Indeed, it is fascinating to note that the measures of the sleep states have, until very recently, all been duration measures. (It is further ironic that as intensity measures have been introduced to the study of sleep, the concept of the states as uniform and stable has broken down.) It should be emphasized that duration is *not* implied by the formal definitions of state given by Ashby (1960); formally, states can be quite instantaneous, but in biological systems relative constancy is the rule. Indeed, we would all agree that we are talking about behavioral conditions that persist for at least one minute when we use the term "state" in a behavioral sense.

While states are relatively enduring in biological systems, it is also true that transitions can occur with dramatic rapidity. That is, there are "jump changes" in state that are virtually discontinuous, and these are particularly helpful to the investigator since they are probably the time of activation of the neural switching processes underlying the behavioral state changes. Thus *Aplysia* can be mobilized from "torpid" to "hungry" in less than a minute, and *Felix* can change from REM sleep to full arousal within five seconds when stimulated. It is interesting to note that transition from activated states (arousal and REM) to inactivated states (torpor and sleep) are always slower than the reverse: hence the ascending limb of the rest-activity cycle is steeper than the descending limb. Perhaps this is why we still know so much less about deactivation processes than about activation processes. An important exception to this generalization is habituation because it is subject to experimental control.

NEURONAL CORRELATES OF BEHAVIORAL STATES AND THEIR INTERPRETATION AS CAUSES

The participants in this symposium all share the conviction that behavioral states have correlated physiological states, including neuronal states. Since the neuron is the functional and structural unit of the nervous system, it is assumed that precise statements about the mechanism and functional significance of states can derive from studies conducted at the neuronal level. When we have established a correlation between a neural state and a behavioral state, we have reason to entertain hypotheses of cause linking the two phenomena. These hypotheses are of three general types, not mutually exclusive:

(1) *The Peripheral Feedback Hypothesis*: The neuronal state is itself a consequence of some aspect of the behavioral state, i.e., the neuron is a sensory neuron receiving an input from an effector of a behavioral component, e.g., an eye movement.

(2) *The Central Feed-Forward Hypothesis*: The observed neuronal state is caused by another as yet unobserved neuronal state which actually generates the behavioral state. It is an incidental rather than a primary cause of the behavioral state.

(3) *The Central Command Hypothesis*: The neural state causes (some aspect of) the behavioral state, i.e., the neuron under study may be a "motoneuron" which drives some effector of a behavioral component, e.g., an eye movement.

An adequate understanding of a state would consist of:

(1) *A complete description of its neural genesis,* i.e., what cells drive what components of the state? The papers in this symposium show strong correlations between neuronal activity and motoric components of behavioral states. In many cases, the neurons are clearly motoneurons or premotor neurons directly or at least tightly coupled to specific state-component events. It is reasonable to construe such correlations as indicative of causes. Vertebrate examples (see Table 3) include the masseteric motoneurons that are inhibited (Chase, 1978) and the reticular giant cells that are excited (McCarley, 1978) in REM sleep,

TABLE 3. Summary of results presented in symposium "Neuronal Correlates of Behavioral State"

Author	Species	CNS locus	Cell system	Method	State	Phenomenon	Observation	Interpretation
Hawkins and Advokat	*Aplysia*	Abdominal ganglion	Sensory and motor neurons	e.c. recording i.c. recording	Arousal	Gill-withdrawal reflex	↑Strength of reflex to tact. stim.	Heterosyn. facil. of sensory neurons, perhaps via cAMP
Chase	*Felix*	Brainstem	Trigeminal motoneurons	i.c. recording	Waking and sleep	Masseteric stretch reflex	Reflex reversed in REM sleep	Depolarization converted to hyperpolarization
Weiss and Kupfermann	*Aplysia*	Cerebral ganglion	Giant serotonergic neurons	e.c. recording i.c. recording	Food arousal	Biting response to food	Increased strength and frequency of biting	Facilitation of motoneurons and muscle contractility, perhaps via cAMP
McCarley	*Felix*	Brainstem	Giant reticular neurons	e.c. recording	Sleep	Spontaneous activity	Increased rate and bursts corr. with REMs	Reciprocal interaction of generator and level-setting neurons
			LC and DRN neurons				Highest rates in W; lowest rates in REM sleep	

producing, respectively, loss of tone in the masseter muscle and rapid eye movements in *Felix*. The invertebrate analogues are the identified motoneurons directly mediating biting (Weiss and Kupfermann, 1978) and gill withdrawal (Hawkins and Advokat, 1978) in *Aplysia*. When we have enough detail of this kind, we can expect to give a complete picture of the execution of the state.

(2) *A complete description of the switching process,* i.e., how are the generator neurons activated and inactivated? Activation of the generator neurons is correlated with events in other central units in ways suggesting that the latter may cause the former. For example, during food arousal in *Aplysia,* the firing of serotonergic cerebral ganglion cells is correlated with the increased strength and frequency of biting (Weiss and Kupfermann, 1978); facilitation of motoneurons and muscle contractility via cyclic AMP may be the direct result of serotonin release (Table 3). The idea that aminergic neurons may thus function as level-setting elements activating (and deactivating) arousal state-component generators has already been suggested by the observation, in mammals, of maximal activity in some locus coeruleus and dorsal raphe nucleus neurons during waking and progressive decreases in the NREM and REM stages of sleep (Steriade and Hobson, 1976; McCarley, 1978).

(3) *An understanding of the functional significance of the state* at the behavioral and neuronal levels. It is intuitively obvious that one function of waking, or food-induced arousal, is to seek, find, capture, and ingest nutrients, and simultaneously, to escape predation — in short, to eat without being eaten. That the same cellular mechanisms may subserve both functions in mammals and invertebrates is an attractive possibility. Sleep or torpor could thus be seen in its traditional energy-conserving, self-preserving context. But REM sleep, with its high levels of central activation and no significant motor output, remains a behavior without an obvious function. Perhaps REM sleep has one function at a behavioral level (that of sleep as adaptive inactivity) and another at the neuronal level (that of internal reorganization).

COMMENTS ON THE PAPERS IN THIS SYMPOSIUM

The subjects of four papers in this symposium are complementary in several important ways:

(1) There is a pair of papers on arousal and sleep (in *Felix*) and another paper on arousal and feeding (in *Aplysia*). These allow the first systematic comparisons of such studies in vertebrates and invertebrates.

(2) One pair of papers studies the neuronal effect of a state change on a classical reflex (one in *Felix*, the other in *Aplysia*). These show that analytically potent intracellular techniques can be applied to both vertebrates and invertebrates under behavioral conditions.

(3) One pair of papers studies the neuronal cause of a state change (arousal in *Aplysia* and REM sleep in *Felix*). These show striking similarities in the working hypotheses emerging from the data.

The methods represent a new level of refinement in behavioral neurobiology. In every case, identified neurons were recorded under precisely specified behavioral conditions. Extracellular and intracellular recordings during state transitions were performed and the electrophysiological data quantitatively analyzed.

State-dependent changes in reflex activity are described and analyzed:

Arousal (compared with quiescence) is characterized by reflex sensitization probably due to heterosynaptic facilitation (Hawkins and Advokat, 1978).

REM sleep (compared to waking and quiet sleep) is characterized by reflex reversal, i.e., suppression instead of facilitation, probably due to monosynaptic inhibition (Chase, 1978).

Indices of a probable role in state mediation are defined and measured (McCarley, 1978; Weiss and Kupfermann, 1978):

Selectivity: The degree to which neuronal firing is state-specific.

Tonic Latency: The activation time or latency of the state-specific firing, with respect to the development of the state.

Phasic Latency: The degree and latency of phase-locked bursts of firing in relation to state-specific movements.

The mechanisms of the state-dependent changes appear to be due to: (1) activation of serotonergic neurons in the arousal and feeding responses (*Aplysia*), producing facilitation of motoneurons and muscle contraction via the second messenger cAMP (Hawkins and Advokat, 1978; Weiss and Kupfermann, 1978); and to (2) inactivation of serotonergic neurons in the REM phase of the mammalian sleep cycle, producing a disinhibition of generator neurons (McCarley, 1978). It thus seems possible that the state-dependent changes in reflex activity (Chase, 1978) could be mediated by disinhibition of inhibitory neurons.

Comparing the results suggests the general hypothesis that *aminergic mechanisms may mediate arousal, perhaps via cAMP, in both vertebrates and invertebrates.* This hypothesis runs diametrically counter to the received ideas about sleep mediation based upon lesion, stimulation, and drug studies. If verified, this hypothesis would serve powerfully to vindicate the Strategy of Simplicity and the single-cell approach, uniting the efforts of those invertebrate and vertebrate neurobiologists who seek to understand the neuronal basis of behavior.

ACKNOWLEDGMENTS

I am grateful to the Commonwealth Fund for its support during my stay at the Institute of Physiology in Pisa, Italy, where this chapter was written. The concepts have grown out of my collaborative work with Dr. Robert W. McCarley, research which has been supported by the National Institute of Mental Health (MH 13,923) and the National Science Foundation.

REFERENCES

Ashby, W. R. (1960). *Design for a Brain*. Chapman and Hull, London.
Bernard, C. (1865). *Introduction à l'Etude de la Medicine Experimentale*. J. B. Bailliere et Fils, Paris.
Chase, M. H. (1978). State-dependent reversal of a brainstem reflex in *Felix domesticus*, pp. 33–65 in *Society for Neuroscience Symposia*, Vol. III, Ferrendelli, J. A., ed. Society for Neuroscience, Bethesda, Md.
Evarts, E. V. (1967). Unit activity in sleep and wakefulness, pp. 545–556 in *The Neurosciences: A Study Program*, Vol. I, Quarton, G. C., T. Melnechuk, and F. O. Schmitt, eds. Rockefeller University Press, New York.
Hawkins, R. D. and C. Advokat (1978). The effects of behavioral state on the gill-withdrawal reflex in *Aplysia californica*, pp. 16–32 in *Society for Neuroscience Symposia*, Vol. III, Ferrendelli, J. A., ed. Society for Neuroscience, Bethesda, Md.
Hebb, D. O. (1955). Drives and the C. N. S. (conceptual nervous system), *Psychol. Rev.* **62**:243–254.
Hinde, R. A. (1966). *Animal Behavior*. McGraw-Hill, London.
Hobson, J. A. (1974). The cellular basis of sleep cycle control, pp. 217–250 in *Advances in Sleep Research*, Vol. I, Weitzman, E. D., ed. Spectrum Publications, Jamaica, NY.
Jouvet, M. (1962). Récherches sur les structures nerveuses et les mecanismes responsables des differentes phases du sommeil physiologique, *Arch. Ital. Biol.* **100**:125–206.
Kandel, E. R. (1976). *Cellular Basis of Behavior*. W. H. Freeman and Co., New York.
Lashley, K. S. (1938). Experimental analysis of instinctive behavior, *Psychol. Rev.* **45**:445–471.

Lorenz, K. (1937). Über die Bildung des Instinktbegriffes, *Naturwissenschaften* **25**:289–300, 307–318, 324–331. In English (Martin, R., trans.) in Lorenz, K., *Studies in Animal and Human Behavior*, Vol. I. Methuen, London, 1970.

McCarley, R. W. (1978). Control of sleep-waking state alteration in *Felix domesticus*, pp. 90–128 in *Society for Neuroscience Symposia*, Vol. III, Ferrendelli, J. A., ed. Society for Neuroscience, Bethesda, Md.

Moruzzi, G. (1969). Sleep and instinctive behavior, *Arch. Ital. Biol.* **107**:175–216.

Ramón y Cajal, S. (1937). *Recollections of My Life*. MIT Press, Cambridge, Mass.

Sherrington, C. S. (1906). *The Integrative Action of the Nervous System*. Yale University Press, New Haven.

Steriade, M. and J. A. Hobson (1976). Neuronal activity during the sleep-waking cycle, *Prog. Neurobiol.* **6**:155–376.

Strumwasser, F. (1971). The cellular basis of behavior in *Aplysia*, *J. Psychiatr. Res.* **8**:237–257.

Tolman, E. C. (1932). *Purposive Behavior in Animals and Men*. Appleton-Century-Crofts, New York.

Weiss, K. R. and I. Kupfermann (1978). Serotonergic neuronal activity and arousal of feeding behavior in *Aplysia californica*, pp. 66–89 in *Society for Neuroscience Symposia*, Vol. III, Ferrendelli, J. A., ed. Society for Neuroscience, Bethesda, Md.

EFFECTS OF BEHAVIORAL STATE ON THE GILL-WITHDRAWAL REFLEX IN *APLYSIA CALIFORNICA*

Robert D. Hawkins and Claire Advokat

*Columbia University College of Physicians and Surgeons,
New York, New York*

INTRODUCTION

Animals are capable of changing their behavior in an adaptive manner in response to changes in the external or internal environment. Such behavioral adaptation is often manifested by alterations in reflex responsiveness. Generally, the alterations in strength of many different reflexes are coordinated in such a way as to achieve an integrated response to the change in the environment. When this occurs, the animal is said to have undergone an alteration in its level of arousal, degree of motivation, or, more generally, "state." Unfortunately, these terms are not well defined, and there is much debate about exactly what they mean and how they should be used. One component they may have in common, however, is sensitization. Sensitization is defined as a change in the strength of a specific reflex (S-R) brought about by presentation of a second stimulus (S'). (These relations are illustrated in Figure 1A.) For instance, S' in arousal could be a loud noise, and in motivation could be a low blood sugar level.

Because it is defined operationally, sensitization may be studied in a precise experimental manner. Furthermore, the sensitization paradigm can be applied to reflex systems that have already proven advantageous for analysis at the cellular level. Sensitization thus seems like an attractive place to start in the investigation of the neural bases of more general phenomena such as arousal and motivation. This paper describes a research program which has been guided by such a strategy.

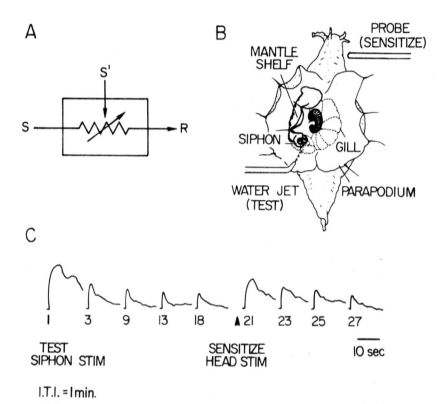

FIGURE 1. Behavioral sensitization. A: Schematic representation of sensitization. In this diagram, the box represents the animal, and the variable resistor represents the strength of a reflex (S-R). Presentation of a second stimulus (S') alters the amplitude or probability of the response (R) evoked by the original stimulus (S). B: Dorsal view of an *Aplysia* prepared for behavioral testing of its gill-withdrawal reflex. The external gill is usually covered by the parapodia, which have been drawn back so that the gill can be seen. A tactile stimulus (water jet) to the siphon causes withdrawal of both the siphon and gill. C: Photocell records of several gill withdrawals evoked by siphon stimulation at 1-min intervals (from Pinsker et al., 1970). A strong stimulus to the head between trials 18 and 21 produces an immediate increase in the amplitude of the response to siphon stimulation.

The first part reviews research on the cellular basis of sensitization of a simple withdrawal reflex, focusing on recent studies carried out by Robert Hawkins, in collaboration with Vincent Castellucci and Eric Kandel. The second part describes behavioral research carried out by Claire Advokat, in collaboration with Tom Carew and Eric Kandel, on the modulation of several defensive reflexes by appetitive stimulation. This research examines the ways in which changes in the strengths of different reflexes are organized and asks what the implications of this behavioral organization might be for the neural organization underlying it.

The specific reflex which has been studied is the defensive gill-withdrawal reflex of the marine mollusc *Aplysia californica*. This animal is advantageous for neuronal analysis because it has a relatively small number of neurons, many of which are large and individually identifiable (Frazier, Kandel, Kupfermann, Waziri, and Coggeshall, 1967; Koester and Kandel, 1977). *Aplysia* also exhibits a variety of simple behaviors, several of which are under behavioral and neurophysiological investigation (for reviews, see Kandel, 1978; Weiss and Kupfermann, 1978).

SENSITIZATION OF THE GILL-WITHDRAWAL REFLEX

Figure 1B shows an *Aplysia* prepared for behavioral testing of its gill-withdrawal reflex. A tactile stimulus to the siphon, which is a fleshy spout at the rear of the mantle, causes withdrawal of both the siphon and gill. If stimulation of the siphon is repeated at regular intervals, the gill withdrawal becomes progressively smaller in amplitude. However, if a noxious stimulus such as prodding or electric shock is applied to the head, the response to siphon stimulation immediately increases in amplitude, as shown in Figure 1C. This is an example of sensitization. Such sensitization may last many minutes (Pinsker, Kupfermann, Castellucci, and Kandel, 1970; Carew, Castellucci, and Kandel, 1971). Repeated noxious stimulation can produce sensitization lasting days (Pinsker, Hening, Carew, and Kandel, 1973).

Figure 2A shows the major elements of the neuronal circuit for the gill-withdrawal reflex (Byrne, Castellucci, and Kandel, 1974; Kupfermann, Carew, and Kandel, 1974). An identified cluster of tactile sensory neurons makes monosynaptic excitatory connections onto six gill motor neurons. Two excitatory interneurons and one inhibitory interneuron have also been identified (Koester and Kandel, 1977). All of these neurons are located in the abdominal ganglion of *Aplysia*.

Kandel and his collaborators have proposed a model of the cellular basis of sensitization of the gill-withdrawal reflex (Kandel, Brunelli, Byrne, and Castellucci, 1976; see Figure 2B). This model has several steps. First, sensitization has been shown to be due to heterosynaptic facilitation at the synapses between the sensory and motor neurons

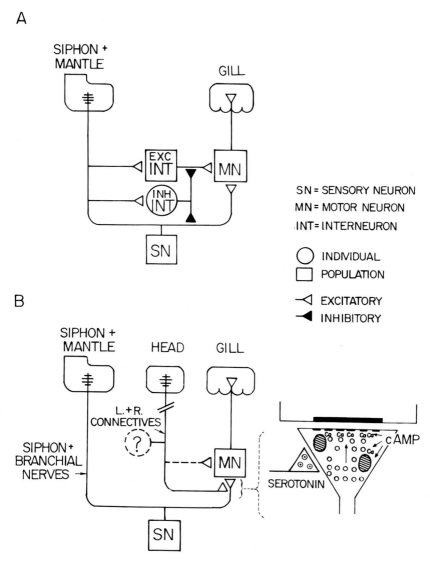

FIGURE 2. Gill-withdrawal circuit and sensitization model. A: Gill-withdrawal circuit. A population of tactile sensory neurons makes monosynaptic excitatory connections onto six gill motor neurons. A few interneurons have also been identified. B: Sensitization model. Sensitization of the gill-withdrawal reflex is due to presynaptic facilitation at the sensory-motor synapses. The facilitating neurons may be serotonergic and may act to increase the levels of cAMP and free Ca in the sensory neuron terminals, leading to greater transmitter release.

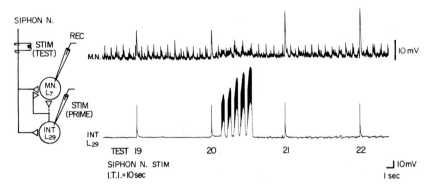

FIGURE 3. Facilitation produced by intracellular stimulation of a single neuron. The experimental arrangement is shown at the left. Top trace: the complex PSP produced in a gill motor neuron by stimulation of the siphon nerve at 10-sec intervals. Bottom trace: intracellular stimulation producing repetitive firing of an interneuron between trials 20 and 21. The PSP in the motor neuron becomes larger after firing of the interneuron.

The other identified facilitating neuron, which will be referred to as L29, is more lateral and indirectly recruits IPSPs onto itself. It seems likely that L29 actually refers to a small population of similar cells, since two facilitating neurons fitting this description have been found in the same ganglion. L29 receives monosynaptic EPSPs from sensory neurons and produces indirect excitation in the gill motor neuron L7. Similarly, many of the unclassified facilitating neurons receive monosynaptic EPSPs from sensory neurons and produce either indirect or monosynaptic EPSPs in L7. These neurons and L29 are thus functionally excitatory interneurons in the circuit for the gill-withdrawal reflex, as well as being modulators of that reflex.

Average Amplitude and Duration of Facilitation

The average facilitation produced by L28, L29, and the unclassified neurons is shown in Figure 4. Recent experiments have concentrated on the two identified neurons, particularly L29, since it has been easier to find and produces facilitation which is both of greater amplitude and longer duration than that produced by L28. The average amplitude of facilitation by L29 is 100%, and the average half-life of the facilitation is about 2 minutes. In some cases the facilitation produced by L29 has lasted as long as 5 minutes.

On the average, the facilitation produced by intracellular stimulation

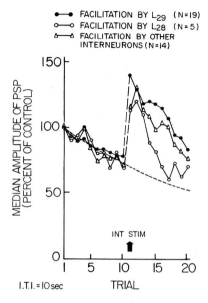

FIGURE 4. Average facilitation produced by intracellular stimulation of a single neuron. Graph of the amplitude of the PSP in the motor neuron on successive trials before and after stimulation of the interneuron (arrow). Open circles: facilitation by L28 (N = 5). Filled circles: facilitation by L29 (N = 19). Triangles: facilitation by other neurons (N = 14).

of a single neuron is smaller in amplitude and shorter in duration than that produced by stimulating one of the connectives, when the two are compared in the same preparations. A possible explanation of this discrepancy is that two or more facilitating neurons are active during connective stimulation. Alternatively, there may be other, more powerful facilitating neurons which have not yet been found.

Facilitation of Monosynaptic EPSP

All the experiments described so far have involved facilitation of the complex PSP produced in a motor neuron by stimulating the siphon or branchial nerves, which contain the axons of the sensory neurons. While this procedure has the advantage of convenience, it is not unambiguous. For instance, such facilitation could be due to homosynaptic facilitation, or PTP, in sensory→interneuron→motor neuron pathways. In order to test this possibility, further experiments were performed on the monosynaptic sensory-motor EPSP. It has been possible to facilitate the monosynaptic PSP by intracellular stimulation of a single facilitating neuron. Since stimulating a facilitating neuron does not produce any firing of the sensory neurons, this facilitation must be heterosynaptic. Moreover, the amplitude and duration of facilitation of the monosynaptic EPSP seem adequate to account for most of the facilitation observed with complex PSPs. More data are necessary to make this comparison quantitatively.

Sensory Neuron Spike Broadening

Klein and Kandel (unpublished data) have recently discovered that connective stimulation, which was known to produce facilitation at the sensory-motor synapses, also produces broadening of the spike in the sensory neuron somas. This broadening is detectable in normal sea water, and is dramatic in tetraethylammonium (TEA) solution. Serotonin and cAMP, both of which produce facilitation, also produce broadening of the sensory neuron spike. Because of these similarities with facilitation of the PSPs at the sensory-motor synapses, it seems plausible that broadening of the spike in the sensory neurons may be related to that facilitation. More specifically, Klein and Kandel propose that in *Aplysia* the soma membrane may serve as a model of the membrane of the synaptic terminals, as has previously been suggested by Stinnakre and Tauc (1973). Changes in the active conductances in the sensory neuron terminals could result in greater Ca influx during a spike and hence in greater transmitter release. The spike broadening is apparently due to an increased Ca conductance, since it is also seen in Na-free solution. The ionic mechanism of this effect is currently under investigation.

If broadening of the spike in sensory neurons is related to facilitation, it should also be produced by stimulation of a single facilitating neuron. Intracellular stimulation of L29 has in fact produced a doubling of the width of the sensory neuron spike in the presence of 60 mM TEA. Thus, all the evidence to date is consistent with the hypothesis that a change in the active conductances in the sensory neuron terminals is a mechanism of facilitation at the sensory-motor synapses.

Neurotransmitter Used by the Facilitating Neurons

Several independent lines of evidence point toward serotonin as being the transmitter which normally produces facilitation at the sensory-motor synapses. Identification of neurons which produce this facilitation has permitted further testing of the serotonin hypothesis. As with facilitation produced by connective stimulation, facilitation produced by inter-neuron stimulation is reversibly reduced by bath application of cinanserin. Cinanserin is known to be a serotonin antagonist in *Aplysia* (Liebeswar, Goldman, Koester, and Mayeri, 1975), although it also affects transmission at some nonserotonergic synapses. This result is then at least consistent with the hypothesis that the facilitating neurons are serotonergic, and has encouraged the undertaking of more definitive biochemical experiments, which are under way. Specifically, we are preparing to apply the enzymatic isotopic assay of Saavedra, Brownstein, and Axelrod (1973) to individual facilitating neurons. This is a sensitive assay for endogenous serotonin which hopefully will provide a more direct answer to the question of whether or not these cells are serotonergic.

The results of the experiments described up to now are summarized schematically in Figure 5. Neurons in the abdominal ganglion have been found which produce the heterosynaptic facilitation underlying behav-ioral sensitization of the gill-withdrawal reflex. These include two identified individuals, one of which is also functionally an excitatory interneuron in the gill-withdrawal circuit. Preliminary results suggest that these neurons may be serotonergic.

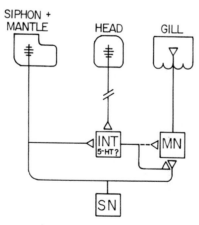

FIGURE 5. Sensitization circuit. Neurons which produce facilitation at the sensory-motor synapses have been found in the abdominal ganglion. Two are identified individuals; one of these is functionally an excitatory interneuron in the circuit for the gill-withdrawal reflex as well as participating in extrinsic modulation of that reflex.

EFFECT OF FEEDING ON DEFENSIVE REFLEXES

The research described so far has followed a reductionist approach, attempting to push the cellular analysis of a particular behavioral phenomenon to greater and greater levels of detail. It is also of interest to take a synthetic approach and attempt to integrate the analyses of several different behavioral phenomena. The goal of this approach is to find the principles of organization which govern the coincident variations in strength or probability of different behaviors, and ultimately to reconstruct the behavior of the organism as a whole. The second part of this paper reviews a series of experiments which have followed this approach.

The specific strategy of this research has been to look at the effect on the defensive withdrawal reflex of another type of stimulation, in this case appetitive food stimulation instead of noxious shock stimulation. Feeding was chosen because (1) it produces a different type of behavior from that elicited by noxious stimuli (that is, appetitive as opposed to defensive), (2) it plays a major role in the animal's behavioral repertoire, (3) it is reasonably well understood behaviorally, and (4) the neural basis of its control is under active investigation (Weiss and Kupfermann, 1978). The goal of the first experiments, then, was to discover what, if any, effect feeding has on defensive withdrawal on a purely behavioral level.

Effect of Feeding on Defensive Withdrawal

The reflex primarily studied has been that of defensive withdrawal of the siphon, because this reflex can be observed in completely unrestrained animals. Furthermore, siphon withdrawal and gill withdrawal in response to moderate tactile stimuli are highly correlated, so one can be used as an index of the other (Carew, Pinsker, and Kandel, 1972).

In these experiments, animals were fed once a day. Presentation of food elicits a complex appetitive response which consists of orienting, head waving, and mouthing. It should be emphasized that after a daily meal the animals still exhibited these food-seeking behaviors—that is, they were not satiated. In fact, they could be said to be in a state of food arousal.

In the first experiment, animals were divided into two groups. The animals in one group were tested immediately after feeding, and those in the other group were tested 24 hours after their last meal. The results of this experiment are shown in Figure 6A. Duration of siphon withdrawal in response to tactile stimulation of the siphon was significantly reduced in the fed animals, as compared to the unfed animals. One week later the two groups were retested with the conditions reversed; that is, the group which previously had been tested 24 hours after feeding was now tested immediately after feeding, and vice versa. Once again, duration of siphon withdrawal was significantly depressed in the fed animals. A second experiment has shown that siphon withdrawal is still significantly depressed 30 minutes after feeding. Additional experiments have shown that these effects can be produced by food stimulation alone, without actual ingestion of food. In experiments currently in progress, the amplitude of gill withdrawal (instead of the duration of siphon withdrawal) has been measured by videotape in restrained animals. These experiments show that gill withdrawal is also depressed following feeding.

Effect of Feeding on Escape and Inking

The experiments described so far have examined the effect of feeding and food stimulation on the defensive withdrawal reflex. It would also be interesting to know if feeding has more general effects on reflex

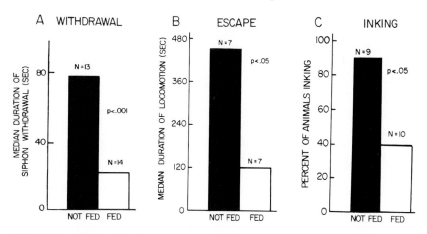

FIGURE 6.   Effect of feeding on defensive reflexes. A: Siphon withdrawal. Median duration of withdrawal in groups of animals that were fed immediately before testing or not fed for 24 h. B: Escape locomotion. Median duration of locomotion evoked by salt crystals on the tail in fed and unfed groups of animals. C: Inking. Percent of animals in fed and unfed groups which inked in response to noxious shock to the neck.

responsiveness, particularly on other defensive reflexes such as escape and inking. Accordingly, two more experiments were performed, the results of which are also shown in Figure 6.

Part B of this figure shows the effect of feeding on escape locomotion. In this experiment, locomotion was elicited by applying salt crystals to the base of the siphon. The duration of locomotion was significantly depressed in animals which had just been fed, as compared to animals which had not been fed for 24 hours.

Part C shows the effect of feeding on inking. If sufficiently provoked, *Aplysia*, like several other molluscs, will release a dense cloud of dark purple ink. Inking is interesting because it tends to be all-or-none, instead of graded like the withdrawal and escape responses (Carew and Kandel, 1977). In this experiment inking was elicited by an electric shock delivered to the neck through previously implanted electrodes. The voltage of stimulation was the same for all animals and was set at the threshold for inking as determined from pilot studies. The percentage of animals which inked in response to this stimulation was significantly lower in the group which had been fed just prior to testing, as compared to the group which had not been fed for 24 hours.

DISCUSSION

Figure 7A summarizes the behavioral evidence to date. All three of the defensive responses tested—gill and siphon withdrawal, escape locomotion, and inking—are depressed in strength or probability by food stimulation. In contrast, gill withdrawal is enhanced or sensitized by noxious stimuli such as electric shock. Conversely, the appetitive biting response is enhanced by food stimulation (Susswein, Kupfermann, and Weiss, 1976), while biting is depressed by strong shock (Kupfermann and Pinsker, 1968). Preston and Lee (1973) have also shown that withdrawal of the head is depressed by food arousal, while approach behavior is enhanced. In these cases, then, examples of two types of behavior (defensive and appetitive) are affected in opposite directions by the same sensitizing stimuli, and each behavior is affected in opposite directions by the two types of stimulation (food and shock).

An organizational scheme suggested by these data is shown in Figure 7B. If one ignores the two boxes labeled "fear" and "food arousal," this diagram is simply a restatement of the behavioral results. The boxes represent the traditional concept of motivational state, that is, intervening variables hypothesized to account for coincident variations

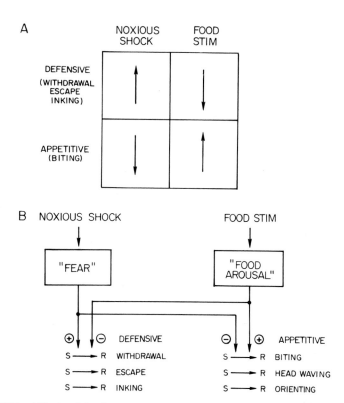

FIGURE 7. Effects of feeding and shock on defensive and appetitive behavior. A: Table summarizing the results of behavioral experiments. Food or shock stimulation alters the strength of defensive and appetitive reflexes. B: Organizational scheme suggested by the results shown in A. The boxes represent hypothesized intervening variables such as motivational state. Changes in state lead to systematic changes in the strengths of related reflexes.

in the strength of many related behaviors. "State" in this sense may simply be a shorthand notation for the behavioral observations—no more and no less—or it may represent a real causal arrangement within the structure of the organism. The latter alternative poses questions which are essentially neurophysiological (as opposed to behavioral) in nature. For instance, is one set of neurons responsible for sensitization of all the behaviors, or are there different sensitizing neurons for each behavior? If there are different sensitizing neurons, how are they interconnected? Are they controlled by other neurons in a hierarchical fashion? If so, what is the structure of the hierarchy? How are the inhibitory effects mediated?

We are now in a position to start investigating these sorts of questions in *Aplysia*. We already know the neural elements which control several of the behaviors listed in this figure, as well as a few neurons which seem to be involved in modulation of those behaviors: L28, L29, and others which are implicated in sensitization of the defensive gill-withdrawal reflex, and the metacerebral cells, which are implicated in sensitization of the appetitive biting response (Weiss and Kupfermann, 1978). Thus, different neurons in different ganglia appear to mediate these two types of sensitization. However, those neurons seem to have interesting biochemical similarities in that they may be serotonergic and produce their effects by increasing the postsynaptic levels of cAMP and free Ca. Thus, although cells mediating sensitization of different behaviors in *Aplysia* may be distributed anatomically, they may use common biochemical mechanisms.

Such generalizations are still very tentative and preliminary. We feel that the research reviewed here, however, justifies some optimism about the continued fruitfulness of this approach. In invertebrates such as *Aplysia* it is possible to identify a few critical neurons controlling a behavior and then to test very specific hypotheses about the mechanisms of that control. Furthermore, different types of behavior in *Aplysia* interact in systematic ways. It should now be possible to study the neural basis of those interactions—that is, the neural basis of behavioral state—with the same sorts of cellular techniques which have been so successful up to now in the analysis of simpler behavioral phenomena in invertebrates.

ACKNOWLEDGMENTS

We would like to thank Kathrin Hilten for preparing the figures, and Vincent Castellucci and Eric Kandel for their comments on the manuscript.

The research reported in this paper was supported by National Institutes of Health postdoctoral fellowships to RDH and CA, and by grants from the Public Health Service (MH 26212) and the McKnight Foundation to Eric Kandel.

REFERENCES

Brunelli, M., V. Castellucci, and E. R. Kandel (1976). Synaptic facilitation and behavioral sensitization in *Aplysia*: possible role of serotonin and cyclic AMP, *Science* **194**:1178–1181.

Byrne, J., V. Castellucci, and E. R. Kandel (1974). Receptive fields and response properties of mechanoreceptor neurons innervating siphon skin and mantle shelf in *Aplysia, J. Neurophysiol.* **37**:1041–1064.

Carew, T. J., V. F. Castellucci, and E. R. Kandel (1971). An analysis of dishabituation and sensitization of the gill-withdrawal reflex in *Aplysia, Int. J. Neurosci.* **2**:79–98.

Carew, T. J. and E. R. Kandel (1977). Inking in *Aplysia californica*. I. Neural circuit of an all-or-none behavioral response, *J. Neurophysiol.* **40**:692–707.

Carew, T. J., H. M. Pinsker, and E. R. Kandel (1972). Long-term habituation of a defensive withdrawal reflex in *Aplysia, Science* **175**:451–454.

Castellucci, V. and E. R. Kandel (1976). Presynaptic facilitation as a mechanism for behavioral sensitization in *Aplysia, Science* **194**:1176–1178.

Castellucci, V., H. Pinsker, I. Kupfermann, and E. R. Kandel (1970). Neuronal mechanisms of habituation and dishabituation of the gill-withdrawal reflex in *Aplysia, Science* **167**:1745–1748.

Cedar, H., E. R. Kandel, and J. H. Schwartz (1972). Cyclic adenosine monophosphate in the nervous system of *Aplysia californica*. I. Increased synthesis in response to synaptic stimulation, *J. Gen. Physiol.* **60**:558–569.

Cedar, H. and J. H. Schwartz (1972). Cyclic adenosine monophosphate in the nervous system of *Aplysia californica*. II. Effect of serotonin and dopamine, *J. Gen. Physiol.* **60**:570–587.

Frazier, W. T., E. R. Kandel, I. Kupfermann, R. Waziri, and R. Coggeshall (1967). Morphological and functional properties of identified neurons in the abdominal ganglion of *Aplysia californica, J. Neurophysiol.* **30**:1288–1351.

Kandel, E. R. (1978). *A Cell-Biological Approach to Learning*, Grass Lecture Monograph 1. Society for Neuroscience, Bethesda, Md.

Kandel, E. R., M. Brunelli, J. Byrne, and V. Castellucci (1976). A common presynaptic locus for the synaptic changes underlying short-term habituation and sensitization of the gill-withdrawal reflex in *Aplysia, Cold Spring Harbor Symp. Quant. Biol.* **40**:465–482.

Koester, J. and E. R. Kandel (1977). Further identification of neurons in the abdominal ganglion of *Aplysia* using behavioral criteria, *Brain Res.* **121**:1–20.

Kupfermann, I., T. J. Carew, and E. R. Kandel (1974). Local, reflex, and central commands controlling gill and siphon movements in *Aplysia, J. Neurophysiol.* **37**:996–1019.

Kupfermann, I. and H. Pinsker (1968). A behavioral modification of the feeding reflex in *Aplysia californica, Commun. Behav. Biol. Part A Orig. Artic.* **2**:13–17.

Levitan, I. B. and S. H. Barondes (1974). Octopamine- and serotonin-stimulated phosphorylation of specific protein in the abdominal ganglion of *Aplysia californica, Proc. Natl. Acad. Sci. U.S.A.* **71**:1145–1148.

Liebeswar, G., J. E. Goldman, J. Koester, and E. Mayeri (1975). Neuronal control of circulation in *Aplysia*. III. Neurotransmitters, *J. Neurophysiol.* **38**:767–779.

Pinsker, H. M., W. A. Hening, T. J. Carew, and E. R. Kandel (1973). Long-term sensitization of the defensive withdrawal reflex in *Aplysia, Science* **182**:1039–1042.

Pinsker, H., I. Kupfermann, V. Castellucci, and E. R. Kandel (1970). Habituation and dishabituation of the gill-withdrawal reflex in *Aplysia*, *Science* **167**:1740–1742.

Preston, R. J. and R. M. Lee (1973). Feeding behavior in *Aplysia californica*: role of chemical and tactile stimuli, *J. Comp. Physiol. Psychol.* **82**:368–381.

Saavedra, J. M., M. Brownstein, and J. Axelrod (1973). A specific and sensitive enzymatic isotopic microassay for serotonin in tissues, *J. Pharmacol. Exp. Ther.* **186**:508–515.

Shimahara, T. and L. Tauc (1977). Cyclic AMP induced by serotonin modulates the activity of an identified synapse in *Aplysia* by facilitating the active permeability to calcium, *Brain Res.* **127**:168–172.

Stinnakre, J. and L. Tauc (1973). Calcium influx in active *Aplysia* neurones detected by injected aequorin, *Nature New Biol.* **242**:113–115.

Susswein, A. J., I. Kupfermann, and K. R. Weiss (1976). The stimulus control of biting in *Aplysia*, *J. Comp. Physiol.* **108**:75–96.

Weiss, K. R. and I. Kupfermann (1978). Serotonergic neuronal activity and arousal of feeding in *Aplysia californica*, pp. 66–89 in *Society for Neuroscience Symposia*, Vol. III, Ferrendelli, J. A., ed. Society for Neuroscience, Bethesda, Md.

STATE-DEPENDENT REVERSAL OF A BRAINSTEM REFLEX IN *FELIX DOMESTICUS*

Michael H. Chase

University of California, Los Angeles, California

INTRODUCTION

The thematic line that our experiments have followed entails the identification of central neural systems that control the behavioral states of sleep and wakefulness and the determination of the mechanisms by which these systems exert their control. We have adopted the position that behavioral states, from a global perspective, may be viewed as specific central nervous system (CNS)-dictated patterns of somato-motor activity. Thus we have felt that the CNS determinants of a behavioral state might be illuminated by studying the CNS control of basic motor processes. In this regard, a clear reflection of general motor control may be found in the tonic excitability fluctuations of simple monosynaptic stretch reflexes, for these basic reflexes form an integral part of the motor background for almost all behaviors. As such, the control of monosynaptic reflex excitability can serve, mirror-like, to reflect the CNS regions, circuitry, and mechanisms that maintain the motor correlates of an animal's behavioral state. Additionally, it is possible that state-dependent patterns of motor control may also mirror the processes which govern the occurrence and maintenance of the states of sleep and wakefulness.

Accordingly, we have reasoned that if a given neural area participates in the motor regulation of a state of sleep or wakefulness, it would, when stimulated, induce a pattern of somatomotor modulation comparable to that which occurs spontaneously during that state. On the other hand, if stimulation initiates a different pattern of somatomotor activity, we would conclude that the area is not involved in maintaining the background level of motor activity of that state, but rather would serve to meet phasic imperatives in response to momentary intrastate directives.

Following this line of reasoning, we conducted a series of experiments on unanesthetized, freely moving cats in order to determine the role of the orbital cortex and brainstem in the control of motor processes during sleep and wakefulness. These two regions were chosen for analysis on the basis of their involvement in sensory, motor, and integrative functions, their participation in state regulation, and their ability to influence somatic reflex activity (Chase and McGinty, 1970a; Jouvet, 1972; Kaada, 1951; Moruzzi, 1972; Rossi and Zanchetti, 1957).

In order to pursue this research strategy, it was first necessary to develop a sensitive test of somatomotor reflex excitability with which to document the effect of conditioning stimulation during sleep and wakefulness. For this purpose we utilized the excitability fluctuations of the monosynaptic masseteric reflex.

As with other monosynaptic reflexes, the factors which control masseteric reflex activity can be analyzed in a relatively straightforward fashion because the reflex response is solely dependent upon the excitability of presynaptic group 1a afferent terminals and postsynaptic motoneurons (Szentagothai, 1948). Moreover, the reflex response, i.e., contraction of the homonymous muscle, is easily recorded as a discrete event time-locked to afferent stimulation. Our reasons for choosing to study the masseteric reflex in particular were based on historical precedent relating to our prior experience and surgical skills, and a general fondness for this particular brainstem reflex. Additionally, we were interested in obtaining information regarding the modulation of reflex excitability at a brainstem level.

With this orientation in mind, I will briefly recapitulate our earlier studies of masseteric reflex excitability during sleep and wakefulness and the state-dependent pattern of reflex control exerted by the orbital cortex and brainstem. Then I will review our recent intracellular studies of the underlying synaptic processes which mediate spontaneous and CNS-induced changes in motoneuron excitability during sleep and wakefulness. A "Model of Motor Control," which is derived from these experiments, will then be presented. The intent of the Model is to provide a reasonable, working hypothesis to account for somatomotor control during the behavioral states of sleep and wakefulness.

EXCITABILITY FLUCTUATIONS OF THE MASSETERIC REFLEX DURING SLEEP AND WAKEFULNESS

The masseteric reflex produces jaw closure and is initiated under natural conditions by the excitation of proprioceptive stretch receptors located within the masseter muscle (McIntyre, 1951). Afferent activity is carried into the brainstem by 1a fibers whose cell bodies lie within the mesencephalic Vth nucleus (Thelander, 1924). These monopolar sensory cells project monosynaptically to neurons in the motor nucleus of the fifth nerve (Szentagothai, 1948). Excitation of these motoneurons then leads to contraction of the masseter muscle. This process is dependent upon synaptic organization located entirely within the brainstem.

The masseteric reflex may also be induced by direct electrical excitation of the sensory cell bodies within the mesencephalic Vth nucleus (Hugelin and Bonvallet, 1957). It is this electrically induced monosynaptic reflex which was utilized in the investigations reported in this chapter. A description of the surgical procedures for establishing this reflex in the freely moving cat and for monitoring its response may be found in the legend of Figure 1 and in reports by Chase (1974); Chase and McGinty (1970*a*); and Chase, McGinty, and Sterman (1968).

Our initial objective was to determine the spontaneous fluctuations in masseteric reflex excitability during the behavioral states of sleep and wakefulness (Chase et al., 1968). The data presented in Figure 2 are the results of a statistical analysis of the amplitude of the masseteric reflex during wakefulness, drowsiness, quiet sleep, and active sleep. The data were obtained utilizing planned comparison tests based upon an analysis of variance. Each sequential change in state (i.e., alert [wakefulness] compared with drowsiness, drowsiness compared with quiet sleep, and quiet sleep compared with active sleep) was marked by a statistically significant reduction in the amplitude of the reflex response (p less than 0.05).[1] The state-dependent fluctuations in amplitude of the reflex contraction of the masseter muscle were consistent between animals and over time.

The description of masseteric reflex excitability during sleep and

---

[1] In our subsequent studies dealing with CNS conditioning stimulation, no qualitative differences in reflex modulation were noted between the waking and drowsy states. For this reason, and to facilitate the collection and analysis of data, the drowsy state was not considered separately in these experiments.

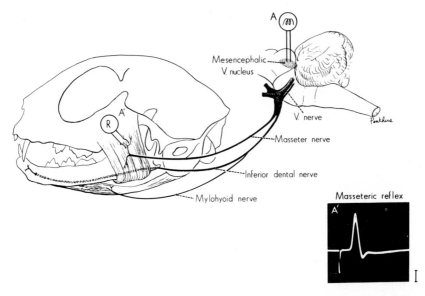

FIGURE 1.   Induced reflex activity within the masseter muscle (A′) was obtained by electrical stimulation of the ipsilateral mesencephalic nucleus of the Vth nerve (A). The bipolar recording electrodes within the masseter muscle are not shown. Calibration line: 10 msec; 200 μV.

wakefulness is therefore straightforward. The mean amplitude of the masseteric reflex is largest during wakefulness; it decreases in amplitude during quiet sleep, while during active sleep reflex excitability is minimal. The relative degree of amplitude depression is clearly greatest when quiet sleep is compared with active sleep. Indeed, the magnitude of depression during active sleep is such that stimulation of the mesencephalic Vth nucleus at multiples of waking threshold values often fails to elicit a response at all (Figure 2D).

A great many studies have demonstrated that spinal cord somatic reflexes exhibit a consistent pattern of state-dependent changes in ex-

FIGURE 2.   Frequency histograms of the amplitudes of 80 consecutive masseteric reflex responses. Reflex activity was obtained during the alert, drowsy, quiet sleep, and active sleep states. The amplitudes of the reflex responses are plotted on an arbitrary scale as a function of the frequency of their occurrence. High-amplitude potentials are reduced and then almost totally abolished as the animal progresses from wakefulness, through drowsiness and quiet sleep, into active sleep. From Chase et al. (1968).

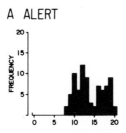

A ALERT

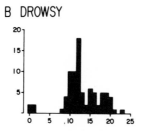

B DROWSY

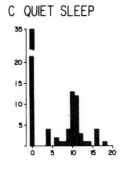

C QUIET SLEEP

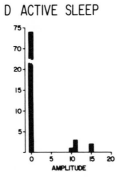

D ACTIVE SLEEP

citability (reviewed in Pompeiano, 1967*a*, *b*). This pattern consists of a reduction in reflex amplitude as the animal passes from wakefulness to active sleep. We have found similar fluctuations in amplitude for the masseteric reflex (see also Enomoto, Fukuoka, Imai, Kako, Kaneko, Mishimagi, Ono, and Kubota, 1968). Thus, on the basis of *prima facie* evidence, we assume that the central neural mechanisms which control spinal cord reflex activity during sleep and wakefulness maintain similar responsibilities in the state-dependent modulation of the brainstem masseteric reflex. Accordingly, we proceed with the belief that the mechanisms determining masseteric reflex excitability reflect motor control processes that are pan-neuraxial in nature. Should this assumption prove to be in error, then the CNS mechanisms of motor control utilized at the level of the brainstem would reflect unique patterns of neuronal organization localized to this region of the neuraxis and thus would be of equally great interest in their own right.

ORBITAL CORTICAL CONTROL OF REFLEX EXCITABILITY DURING SLEEP AND WAKEFULNESS

Tonic masseteric reflex excitability is normally greatest during wake-fulness, less during quiet sleep, and least during active sleep (Chase et al., 1968). Utilizing this data as a control, the effect of unilateral stimulation of the orbital cortex on ipsilateral masseteric reflex excita-bility was examined in freely moving cats (Chase and McGinty, 1970*a*, *b*).

During wakefulness, orbital cortical stimulation resulted in strong suppression of the electromyographically recorded masseteric reflex (Figure 3). The relative degree of suppression was reduced during quiet sleep (Figure 4). During active sleep, only a very slight degree of reflex suppression was observed with the identical stimulus that previously had been employed during the waking and quiet sleep states (Figure 4). Thus, as the animal passed from wakefulness to quiet sleep to active sleep, the degree to which orbital stimulation was capable of inducing reflex suppression diminished.

The decreasing effectiveness of the orbital cortex in producing reflex suppression during quiet and active sleep was examined with a number of different paradigms, including (1) fixed parameters of reflex and or-bital excitation, (2) fixed parameters of reflex induction and variable levels of orbital stimulation, and (3) variable parameters of reflex induction and fixed levels of orbital stimulation (Chase and McGinty, 1970*b*).

a. REFLEX-CONTROL   b. CORTICAL STIMULATION

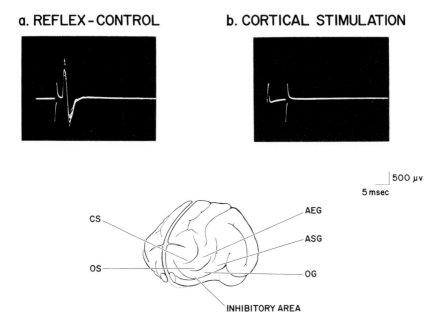

500 μv

5 msec

FIGURE 3.   This figure illustrates the orbital cortical area whose excitation results in potent suppression of the electromyographically recorded masseteric reflex during wakefulness in the freely moving cat. The oscilloscope records demonstrate the potency of this effect during the waking state, for a single conditioning pulse delivered to the orbital cortex led to complete reflex suppression. This same area was found to induce postsynaptic inhibition of jaw-closer neurons in acute intracellular experiments (Nakamura et al., 1967). Mesencephalic Vth nucleus: 4 V, 0.8 msec; orbital cortex: 8 V, 0.8 msec. Abbreviations: CS, coronal sulcus; OS, orbital sulcus; OG, orbital gyrus; ASG, anterior sylvian gyrus; AEG, anterior ectosylvian gyrus.

In all of these studies the same state-dependent pattern of reflex modulation was observed. Reflex suppression was most pronounced during wakefulness. The degree of suppression decreased during quiet sleep and became minimal during active sleep.

Since orbital control of reflex excitability was most prominent during wakefulness and least effective during active sleep, we conclude that the capability of the orbital cortex for the production of motor suppression is, in a sense, "turned on" during wakefulness and "turned off" during active sleep.

During wakefulness, the functional result of orbital cortical activation is to produce a qualitative departure from the tonically maintained

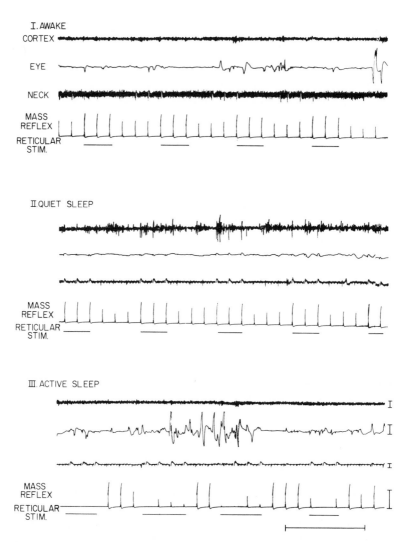

FIGURE 5. Polygraphic recordings of the amplitude of the masseteric reflex and the activity of the sensorimotor cortex, eyes, and neck musculature. The reflex was recorded, on-line, by employing a window circuit to isolate the reflex and a peak-amplitude, pulse-lengthening circuit to deflect the polygraphic pen commensurate with the reflex amplitude. Note the consistency of the reticular effect between states and within states, especially during periods of phasic activity such as ocular movements and spindle bursts. Mesencephalic Vth nucleus: (I) 5 V, 0.5 msec; (II) 6 V, 0.5 msec; (III) 7 V, 0.5 msec. Reticular tegmentum: 5 V, 0.9 msec, 3 pulses. Calibration: cortex, eye, neck, 50 μV; reflex, 500 μV; time, 5 sec. From Chase and Babb (1973).

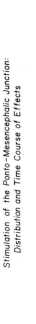

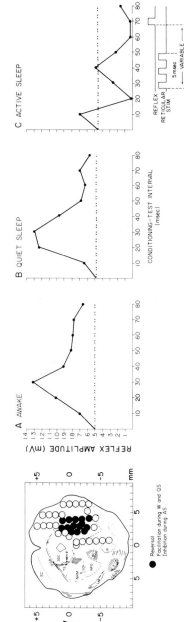

FIGURE 6.   The solid circles in the brainstem cross-section indicate the sites which yielded reticular-conditioned masseteric reflex facilitation during wakefulness and quiet sleep, and suppression during active sleep. The typical time course of this pattern of response is shown in A–C. Note the potent long-duration facilitatory effect during wakefulness and quiet sleep (A, B) and the biphasic period of suppression during active sleep (C). We occasionally observed sustained suppression during active sleep when the return to baseline of the response at 40 msec was absent. In most animals, a brief period of low-amplitude facilitation persisted during active sleep at a conditioning-test latency of 5 to 10 msec. All of the sites (empty circles as well as solid circles) shown in this brainstem section induced a pattern of suppression during active sleep similar to that shown in A–C. This response was observed even when no effect was found during wakefulness or quiet sleep. From Wills and Chase (1978).

correlate this data with the results of single pulse and short pulse-train conditioning stimuli. For this series of experiments the masseteric reflex was induced continuously at a rate of 1.5 per sec. The reticular conditioning site was stimulated at random intervals (none of which was less than 3 min) for periods of 4 sec at a frequency of 100 cycles per sec. The average amplitude of the six reflex responses induced prior to reticular stimulation served as the control for the six reflex responses which were induced during the conditioning stimulus.

High-frequency reticular conditioning stimulation produced a pattern of reflex facilitation during wakefulness and quiet sleep and suppression during active sleep (Figures 7 and 8). This state-dependent pattern was

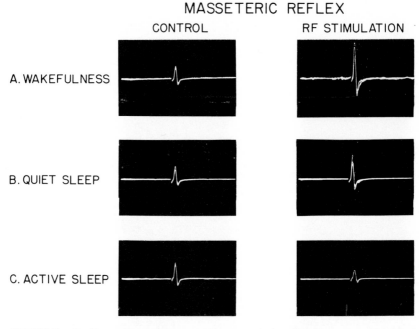

FIGURE 7.   Oscilloscopic recordings of the masseteric reflex (three superimposed traces per photograph). The level of reflex induction was increased from wakefulness to quiet sleep to active sleep in order to maintain equivalent control amplitudes (A: 3 V, 0.5 msec; B: 4 V, 0.5 msec; C: 5.5 V, 0.5 msec). The level of reticular stimulation was held constant throughout all states (4 V, 0.75 msec, 100 pulses/sec). Note that no masseteric muscular response occurred in conjunction with reticular stimulation. Reflex facilitation was induced by reticular excitation during wakefulness (A) and quiet sleep (B), while suppression resulted during active sleep (C). Calibration: 5 msec, 300 μV. From Chase et al. (1976).

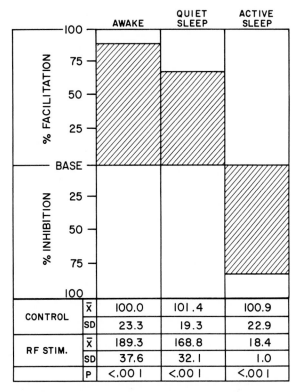

FIGURE 8.   Graphic and statistical description of the masseteric reflex response to reticular stimulation. For this study the frequency of reflex induction was 1.5/sec, and the level of reflex induction for the awake, quiet sleep, and active sleep states were, respectively, 3 V, 0.5 msec; 3.5 V, 0.5 msec; and 4.2 V, 0.5 msec. Reticular stimulation (4 V, 0.5 msec, 100 pulses/sec for 4 sec) was held constant. The percent variation in reflex amplitude induced by reticular stimulation (plotted on an arbitrary but relative scale) was obtained from data collected from 100 separate trials during each state (each trial consisted of a comparison of six control reflexes compared with six reflexes modified by reticular stimulation). From Chase et al. (1976).

observed consistently throughout consecutive states of wakefulness, quiet sleep, and active sleep (Figure 8). No qualitative variations occurred during intrastate epochs such as those encompassed by the eye movement periods of wakefulness or active sleep, or the spindle activity of quiet sleep.

While the direction of the reflex response remained constant throughout each state, the degree of suppression and facilitation depended upon the parameters of the conditioning stimulus. The effectiveness of reticular stimulation (in promoting either facilitation or suppression) became greater as the degree of reticular activation was augmented (by increasing the voltage, number, or duration of the conditioning pulses).

In summary, a consistent and unvarying pattern of reflex modulation was elicited by single pulse, pulse-train, or high-frequency reticular stimulation. When the animal was awake or in quiet sleep, conditioning stimulation was accompanied by reflex facilitation. As soon as the animal entered the active sleep state, the identical conditioning stimulus produced reflex suppression. This phenomenon, which we have termed reticular "response-reversal," is most dramatically evident when one is observing a freely moving cat during a particular behavioral state. For example, during active sleep, with the conditioning-test paradigm or during high-frequency conditioning trials, the test reflex can be completely suppressed by reticular stimulation. At the very moment that the animal either awakens or passes into quiet sleep, the test reflex becomes strongly facilitated, with an amplitude often double that of its control, and all without the intervention of the investigator or any change in stimulating or recording parameters. It is the animal's state alone that "determines" whether the conditioned response is one of motor suppression or facilitation.

In other experiments which have just been completed and are in the process of being prepared for publication, we have examined the effect of conditioning stimulation delivered to rostral sites at the mesodiencephalic junction and caudal sites at the pontomedullary junction (Chase and Wills, 1978; Wills and Chase, 1978; for preliminary reports see Wills-Lewis and Chase, 1976; and Babb, Wills-Lewis, and Chase, 1976). When the animal was awake or in quiet sleep, reticular stimulation led either to reflex facilitation (at most conditioning-test intervals) or was ineffective in promoting any change in reflex excitability. However, during active sleep, from all sites at these levels of the neuraxis (as well as at the pontomesencephalic level), reflex suppression was generated with a time course closely approximating that shown in Figure 6. At those sites which induced reflex facilitation during either wakefulness or quiet sleep, reflex suppression during active sleep replaced the facilitatory response. At all sites, reflex suppression was generated in active sleep at a substantially lower threshold than reflex facilitation during wakefulness or quiet sleep.

Thus, it appears that there is a functional reorganization of the entire parenchyma of the brainstem during active sleep. The result is that all regions examined promote a profound suppression of somatomotor reflex activity at extremely low conditioning thresholds. Brainstem stimulation which induces reflex facilitation during waking or quiet sleep instead generates reflex suppression during active sleep. Those sites having neither a waking nor a quiet sleep reflex modulating action gain the potential for reflex suppression as soon as the animal's state changes to active sleep.

While this apparent functional reorganization of motor control within the entire brainstem is intriguing, we have directed our current efforts toward elucidating the mechanisms which underlie the phenomenon of pontomesencephalic reticular response-reversal and toward determining whether comparable processes control the tonic level of motor activity during sleep and wakefulness. This strategy directs our attention to the investigation of the paradoxical finding that brainstem sites apparently can change their motor signature from facilitation to suppression, or from suppression to facilitation, in a manner bound exclusively to the animal's state. It also allows us to study a paradoxical pattern of motor modulation which, as indicated in the preceding paragraph, apparently reflects aspects of the motor control mechanisms utilized by almost all other brainstem sites.

INTRACELLULAR ANALYSIS OF TRIGEMINAL MOTONEURON ACTIVITY DURING SLEEP AND WAKEFULNESS

Having found a pontomesencephalic region capable of promoting a pattern of motor control similar to that which is maintained normally during sleep and wakefulness, we felt that the elucidation of the responsible mechanisms and the determination of their similarity to tonic motor control processes would require the application of intracellular recording methodologies. By such means we would hope to compare the spontaneous state-dependent synaptic activity operating upon masseter motoneurons with that induced by brainstem stimulation.

We felt that a determination of cellular mechanisms would be essential, for the data which could be obtained by the extracellular procedures currently utilized in freely moving animals (for example, evoked potential or extracellular unit recordings) only allude to the synaptic drives controlling neuronal discharge. By recording intracellularly, we would be able to investigate whether motor neuronal discharge was due to the advent of excitatory postsynaptic potentials (EPSPs) or to the with-drawal of inhibition, i.e., disinhibition. On the other hand, decreased motoneuron excitability might result from a process of postsynaptic inhibition or disfacilitation.

Thus, whenever there is cellular activation, it may reflect one of two distinct and opposite patterns of neuronal control; the suppression of activity may likewise reflect separate and opposite patterns of synaptic modulation. We felt that direct access to these mechanisms might be achieved by an intracellular examination of the synaptic events in the motoneuron's life during the animal's behavioral states. By monitoring the membrane potential of individual trigeminal motoneurons, we hoped to achieve a strategic position to view the basic fundamental mecha-nisms and synaptology responsible for motor control during the states of sleep and wakefulness.

Unfortunately, no methods were available to permit an intracellular analysis in the chronic cat during the sleep states (Steriade and Hobson, 1976). Therefore, in the pursuit of the preceding objectives, we de-veloped a technique to record intracellularly in cats that were intact and unanesthetized, respiring normally, and moving freely (with the exception of head fixation). We have held identified trigeminal moto-neurons for extended periods of time throughout contiguous states of wakefulness, quiet sleep, and active sleep.

Our initial intracellular studies involved primarily an examination of the spontaneous changes in membrane polarization of trigeminal moto-neurons during quiet and active sleep; we have also begun an explora-tion of the cellular basis for reticular response-reversal (Nakamura, Goldberg, Chandler, and Chase, 1978). We have returned then, in these experiments, to study the cellular correlates of the electromyographic changes in tonic and induced somatomotor activity that we reported previously (Chase, 1974; Chase, 1976).

The basic experimental paradigm entailed the implantation of chronic stimulating and recording electrodes while the animals were anesthe-tized. The implantation procedures were identical to those utilized in the experiments on freely moving cats described in the preceding sections. In preparation for subsequent intracellular explorations, a small circle (6 mm diameter) of occipital bone was removed by trephina-tion, and the hole was filled with bone wax (which was later extracted to permit the subsequent penetration by a microelectrode through the cerebellum to trigeminal motoneurons). In addition, while the animal's head was stereotaxically positioned, four receptacles designed to receive stereotaxically mounted calibrated bars were fixed to the cal-varium with dental cement. During experimental sessions, by inserting bars into the receptacles and attaching the bars to the stereotaxic instrument according to the calibration determined at the time of im-plantation, the animal's head could be rigidly held in a stereotaxically defined position without any pain or pressure.

After each animal had recovered from surgery (approximately 1 week), it was fixed in the stereotaxic apparatus by means of the head holder described above. The bone wax was removed from the trephined hole overlying the cerebellum and a glass micropipette was lowered through the cerebellum to the motor nucleus of the fifth nerve. Tri-geminal motoneurons which innervate the jaw-closing muscles (jaw-closer motoneurons) were identified by the following criteria. (1) They were located in a region estimated stereotaxically to be the dorsal part of the trigeminal motor nucleus, which contains motoneurons inner-vating the jaw-closing masseter and temporal muscles (Batini, Buis-seret-Delma, and Corvisier, 1976; Landgren and Olsson, 1976; Mizuno, Konishi, and Sato, 1975). In this region, which is approximately 16–18 mm below the cerebellar surface, stimulation of the ipsilateral tri-geminal mesencephalic nucleus evoked an extracellularly recorded monosynaptic negative field potential with a latency of approximately 0.6 msec and an amplitude of 2–6 mV (Figure 9C). (2) Upon penetration of a motoneuron, intracellularly monitored monosynaptic EPSPs with latencies of 0.5 to 0.7 msec were induced by stimulation of the mesencephalic Vth nucleus, thus providing concrete cellular identifica-tion (Figure 9D; Kidokoro, Kubota, Shuto, and Sumino, 1968; Naka-mura, 1978). For all studies, following each cellular penetration, an extracellular record was obtained in the same region during quiet and active sleep. Intracellular data were recorded polygraphically and on magnetic tape, along with EEG, EOG, and EMG activity.

During experimental sessions, after penetration and identification of a trigeminal jaw-closer motoneuron with a micropipette filled with 3 M KCl or 2 M K-citrate (tip resistances, 8 to 20 megohms), the mem-brane potential of the motoneuron was correlated with polygraphic data indicating the state of the animal as one of quiet or active sleep. For certain cells, a single pulse or a short train of two to four pulses (interpulse interval, 2 msec) was delivered to the nucleus reticularis pontis oralis in order to examine the intracellular response of the motoneuron to reticular stimulation. The basic stimulation and recording paradigm is outlined diagrammatically in Figure 9.

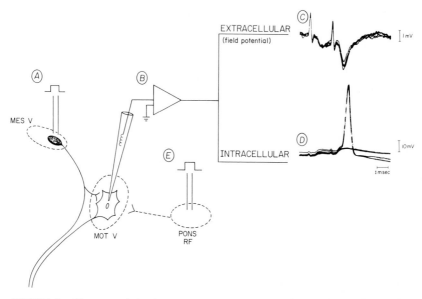

FIGURE 9. Diagram of the basic stimulation and recording paradigm for intracellular studies. The trigeminal motor nucleus (Mot V) was identified by monitoring
its extracellular field potential (C) induced by stimulation (A) of the mesencephalic
nucleus of the fifth nerve (Mes V). Jaw-closer motoneurons were identified by
intracellular recording of the monosynaptic EPSPs and spike potentials of mesencephalic V origin (A, B, D). E: Stimulation of the nucleus pontis oralis (Pons RF).
From Nakamura et al. (1978).

Membrane Potential Changes During Quiet and Active Sleep

The basis for the spontaneous modulation of the masseteric reflex
during sleep and wakefulness was examined by directly monitoring the
level of membrane polarization of jaw-closer motoneurons. As the animal
passed from quiet to active sleep, the membrane potential gradually
became increasingly polarized (Figures 10C and 12C). A tonic hyperpolarization (relative to quiet sleep) of 3 to 10 mV was maintained
throughout the active sleep period (Figure 10C). Phasic periods of
hyperpolarization, superimposed upon the tonic level of hyperpolarization, were also observed during active sleep (Figure 11D).

In this chronic preparation, spontaneous subthreshold synaptic
potentials were also recorded in jaw-closer motoneurons.[2] (See Naka-

---

[2] These potentials were similar in waveform to those described in spinal motoneurons
of the anaemically narcotized cat in conjunction with the stretching of homonymous
and antagonistic muscles (Granit, 1970).

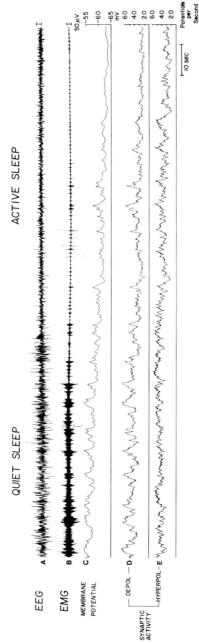

FIGURE 10. Development of motoneuron hyperpolarization as the animal passes from quiet to active sleep. Note the gradual increase in membrane polarization from a quiet sleep level of approximately −55 mV to −64 mV during active sleep. The *frequency* of subthreshold synaptic potentials in the depolarizing (D) and hyperpolarizing (E) direction was less during active sleep than during quiet sleep. This decrease in synaptic activity in active sleep paralleled the increase in tonic hyperpolarization observed during this state. From Nakamura et al. (1978).

in general, the pattern of short-latency, low-amplitude reflex facilitation induced by reticular conditioning stimulation during this state, as described in the preceding section of this chapter.

The profound motoneuron membrane hyperpolarization we observed during active sleep provides a final common pathway criterion for the muscular atonia described by Jouvet, Michael, and Courjon in 1960, which is a characteristic of this state. It also presents direct evidence for the inference that motoneuron hyperpolarization is the prevailing process responsible for tonic motor suppression during active sleep (Pompeiano, 1967a, b).

In addition to an increase in motoneuron polarization during active sleep, we found a rather striking tonic decrease in the frequency of occurrence of subthreshold synaptic potentials. If we assume that subthreshold synaptic activity can be used as a determinant of synaptic bombardment in the region of the motoneuron soma (Burke, 1967; Granit, Kellerth, and Williams, 1964a, b), then at least three possible explanations emerge that could account for the decrease in subthreshold synaptic activity during active sleep. The first possibility entails the onset of a disfacilitatory process which could result in a decrease in excitatory synaptic bombardment as recorded in the region of the neuronal soma. A second explanation involves an increase in inhibitory dendritic synaptic bombardment which could shunt dendritic EPSPs and result in an apparent decrease in synaptic activity recorded in the soma (Kuno, 1971). It is also possible that an increase in very low-level somatic inhibitory input, by impinging on the soma more asynchronously in active sleep than in quiet sleep, might not summate and be observed as subthreshold synaptic potentials. But such an increase in somatic inhibitory synaptic activity is unlikely, since comparable sustained asynchronous input (e.g., natural stimulation of the tongue) results in an increase in subthreshold synaptic potentials of jaw-closer motoneurons (Goldberg, 1968). This activity is induced predominantly via inhibitory somatic synapses and is accompanied as well by tonic membrane hyperpolarization (Goldberg, 1968). The decrease in subthreshold synaptic activity during active sleep therefore suggests that the tonic hyperpolarization which occurs during this state may likely be achieved by the withdrawal of somatic EPSPs and/or by the onset of dendritic IPSPs.[3]

The findings dealing with reticular conditioning stimulation suggest

---

[3] We have relied upon classically defined synaptic mechanisms for our *initial* hypotheses rather than invoking some of the recent concepts of synaptic control exercised by decreased ionic conductances resulting in slow potential activity. In the future we hope to examine these newer data closely to determine if they are relevant to our findings.

that the previously reported reticular-induced reflex facilitation of quiet sleep is due to the advent of a depolarizing potential. We have also found that a comparable depolarizing potential is induced during wakefulness (Chandler, Nakamura, and Chase, 1977). On the other hand, during active sleep, the predominant synaptic drive produces a hyperpolarizing potential, which accounts for the electromyographically recorded reflex suppression observed during this state. Furthermore, the short-latency, low-level reflex facilitatory response recorded electromyographically during active sleep has its counterpart in the short-latency, low-level depolarizing potential. It is clear that this pattern of membrane potential modulation represents the basis for the paradoxical phenomenon of reticular response-reversal.

A MODEL OF MOTOR CONTROL

We began our exploration of the central neural systems that control motor behavior by determining the spontaneous state-dependent variations in amplitude of the masseteric reflex. We found that the reflex was largest (facilitated?) during wakefulness and smallest (inhibited?) during active sleep. We then found that orbital cortical stimulation led to suppression of the reflex response during all states, but that its effect was strongest during wakefulness and weakest during active sleep. On the other hand, excitation of pontomesencephalic reticular sites facilitated the masseteric reflex during wakefulness *and* inhibited it during active sleep, which duplicates the pattern of motor control that occurs spontaneously during these states. Since cortical inhibition is practically ineffective during active sleep, when a maintained pattern of reflex suppression occurs, the orbital "effect" represents a departure in both direction (during wakefulness) and degree (during active sleep) from the spontaneous, tonic state-dependent level of motor excitability. Conversely, reticular stimulation induces activity which mimics the patterns of motor control which normally occur during wakefulness and active sleep.

In order to fit the preceding observations into an integrated scheme, a Model of Motor Control was developed. It essentially accounts for two principal observations: (1) decreasing cortical control during active sleep, and (2) reticular response-reversal during active sleep compared with wakefulness.

The basic thematic concepts of the Model are summarized below.[4]

---

[4] For the present, the quiet sleep state has been omitted from the Model, but in most cases it accords, in our thinking, with the pattern of modulation exercised during the waking state, the major distinction being quantitative, not qualitative.

During Wakefulness

Observation 1: The pontomesencephalic reticular formation initiates EEG desynchronization.

Mechanism: Presumably by processes that result in the activation of cerebral cortical neurons.

Observation 2: Orbitocortical activation is very effective in promoting masseteric reflex suppression.

Mechanism: Presumably due to activation of bulbar inhibitory neurons.

Observation 3: The pontomesencephalic reticular formation induces sustained masseteric reflex *facilitation*.

Mechanism: Presumably by a reticulotrigeminal route.

Supposition: Pontomesencephalic activation of the bulbar inhibitory area is minimal, since these two regions are functionally uncoupled during this state.

Resultant Effects: These factors result in strong orbital cortical suppression of the masseteric reflex, and spontaneous and pontomesencephalic reticular-induced EEG desynchronization and reflex facilitation.

During Active Sleep

Observation 1: The pontomesencephalic reticular formation initiates EEG desynchronization.

Mechanism: Presumably by processes that result in the activation of cerebral cortical neurons.

Observation 2: Orbitocortical activation is relatively ineffective in promoting masseteric reflex suppression.

Mechanism: Presumably due to decreased responsiveness of bulbar inhibitory neurons.

Observation 3: The pontomesencephalic reticular formation induces (a brief period of reflex facilitation that is followed by) sustained masseteric reflex *suppression*.

Mechanism: Presumably by a reticulotrigeminal route for the facilitatory response and by activation of bulbar inhibitory neurons for the suppressor response.

Supposition: Pontomesencephalic activation of the bulbar inhibitory area is established by a coupling mechanism which establishes functional excitatory patency from the rostral region to the caudal region.

Resultant Effects: These factors result in weak orbital cortical inhibition of the masseteric reflex, and spontaneous and pontomesencephalic reticular-induced EEG desynchronization and reflex suppression.

Observation 1: The pontomesencephalic reticular formation activates the cerebral cortex, resulting in EEG desynchronization during wakefulness and active sleep.

There are few manipulations of the CNS that have proven as reliable and repeatable as cortical desynchronization resulting from stimulation of the pontomesencephalic reticular formation. Finely graded electrical stimulation of the reticular formation leads, with almost perfect regularity, to correlated degrees of low-voltage, high-frequency activity over the entire cortical mantle (Bremer, 1961; Dell, 1963; Moruzzi and Magoun, 1949). The maintenance of a desynchronized EEG pattern during wakefulness, in conjunction with strong excitation of the "reticular activating system," was also observed in our experiments. Since the EEG appears maximally desynchronized during active sleep, no specific effect of reticular stimulation was noted, nor did stimulation disrupt the state itself. Therefore, the discharge of neurons of the ponto-mesencephalon either participates in maintaining cortical desynchronization during wakefulness and active sleep, or at the very least is compatible with and does not disrupt the pattern of EEG activity which accompanies these states.

Observation 2: During wakefulness, orbitocortical activation of the bulbar inhibitory area leads to suppression of the "masseteric reflex," but this response is markedly reduced during active sleep.

The data indicating (1) that there are specific orbital cortical effects exerted upon the bulbar reticular formation, and (2) that these influences are maximal during wakefulness and depressed during active sleep begin with a description of the mechanisms of action and the presumptive projection systems.

In acute immobilized cats, Nakamura, Goldberg, and Clemente (1967) studied the nature of the suppression of the masseteric reflex following stimulation of the orbital cortex. Utilizing intracellular techniques, they found that orbital stimulation induced a prolonged hyperpolarizing potential in masseter motoneurons. Further studies led them to conclude that the observed suppression of the masseteric reflex was due to active postsynaptic inhibition. The existence of a specific pathway mediating orbitofugal inhibition of the masseteric reflex has been demonstrated electrophysiologically by Clemente, Chase, Knauss, Sauerland, and Sterman (1966) and Sauerland, Nakamura, and Clemente (1967), and anatomically by Mizuno, Sauerland, and Clemente (1969). In 1966, Clemente et al. reported that short pulse-train electrical stimulation of the orbital cortex, in flaxedilized cats, produced profound suppression of the masseteric reflex. Following this observation, Sauerland et al. (1967), utilizing a comparable preparation, found that the orbital cortex sent direct projections to the ventromedial bulbar reticular formation (chiefly to the nucleus reticularis giganto-cellularis). These experiments demonstrated that single pulse excitation of the anterior portion of the orbital cortex evoked a monosynaptic response in ipsilateral and contralateral bulbar reticular sites. The fact that the ventromedial bulbar reticular formation serves as a relay for orbital-induced inhibition of the monosynaptic masseteric reflex was subsequently demonstrated by transection experiments (Sauerland et al., 1967). In these studies the inhibitory response was lost following isolation of the medullary region by transection of the brainstem at the pontomedullary junction.

The anatomical basis for orbital inhibition of the masseteric reflex was then explored by Mizuno et al. (1969), who reported that "an anatomi-cal substrate exists for the cortically induced influences on reflexes described electrophysiologically by our group." They concluded, on the basis of preterminal degeneration following cortical lesions, that inhibition of the masseteric reflex by orbital stimulation was mediated by a cortico-reticulo-trigeminal pathway.

Our studies in freely moving cats support these electrophysiological and anatomical investigations. However, the basis for our finding that orbital cortical influences are maximally effective during wakefulness and minimally effective during active sleep must await future experi-mentation for clarification. We hypothesize that the effect may be due to a decreased efficacy of corticofugal discharge for activating bulbar neurons during active sleep. The responsible mechanism(s) may also operate at a number of sites, for the depression of the orbital effect may reflect changes of excitability along the entire pathway from cor-tex to motoneuron, including the orbital cortex, a subcortical relay or relays, and the motoneuron itself.

Unfortunately, the cortical control of motor activity during sleep has been examined by only a very few investigators, although one of the earliest studies was carried out in 1894 when Tarchanoff observed that limb movements evoked by motor cortical stimulation were re-duced during sleep. Seventy-one years later, Hodes and Suzuki (1965) applied high-frequency electrical stimulation to the "frontal cortex" of cats and determined that the threshold voltage for inducing head and forelimb movements was lowest in the waking animal and highest during quiet sleep, with an intermediate value for active sleep. Iwama and Kawamoto (1966) also stimulated the frontal (motor) cortex of cats and recorded the activity of flexor hindlimb muscles following short pulse-train stimulation. They found a gradual decrease in the somatic excitatory response as the animal passed from wakefulness through quiet sleep into active sleep. Similarly, Baldissera, Ettore, Infuso, Mancia, and Pagni (1966) reported that motor cortical stimulation produced the largest electromyographic response during wakefulness, whereas during active sleep it was almost absent. Thus, while we observed a reduction in cortical control during active sleep (Chase and McGinty, 1970*a, b*), as did Iwama and Kawamoto (1966) and Baldissera et al. (1966), Hodes and Suzuki (1965) found enhanced corticofugal effects during this state.

Observation 3: The pontomesencephalic reticular formation induces sustained masseteric reflex facilitation during wakefulness, and suppression during active sleep.

As detailed in previous sections of this chapter, reticular stimulation induced changes in the masseteric reflex in the same direction as the spontaneous amplitude variations which occur during wakefulness (facilitation) and active sleep (suppression). Specifically, the reflex was facilitated during wakefulness (and quiet sleep) when reticular formation stimulation preceded the induction of the masseteric reflex (conditioning-test paradigm) or was concurrent with it (high-frequency conditioning paradigm). During active sleep, an initial brief period of weak reflex facilitation was observed, followed by intense and prolonged reflex suppression. The primary effect during active sleep was therefore one of a decrease in reflex excitability, as reflected by its duration and degree, as well as by sustained reflex suppression which accompanied continuous high-frequency reticular stimulation.

Corresponding intracellular data demonstrated the presence, during active sleep, of a potent reticular-induced hyperpolarizing potential which appeared to supersede the depolarizing potential generated during quiet sleep and wakefulness (Chandler et al., 1977). However, an initial and very brief phase of depolarization persisted during active sleep. Thus the intracellular data provide precise motoneuron membrane correlates for the pattern of reticular response-reversal originally observed by extracellularly monitoring the electromyographic contraction of the masseter muscle.

In a series of extracellular acute experiments on immobilized cats, Dell, Bonvallet, and Hugelin found that stimulation of the pontomesencephalic reticular formation led to facilitation of the masseteric reflex (Dell, 1963; Dell, Bonvallet, and Hugelin, 1961; Hugelin, 1961; Hugelin and Bonvallet, 1957). In addition, Sauerland et al. (1967) reported that excitation of the pontine reticular formation resulted in masseteric reflex facilitation. In this latter study it was shown that bulbar structures were not involved in the facilitatory response and that ponto-mesencephalic sites were responsible. Since the studies of both groups were carried out in the immobilized preparation (under flaxedil and/or urethane), the state of the animal was unknown, but was presumably wakefulness.

It is clear from our own studies as well as those cited above that ponto-mesencephalic reticular stimulation facilitates the masseteric reflex in the waking cat, probably by a reticulotrigeminal route (Sauerland et al., 1967). However, we have demonstrated electromyographically a pattern of sustained reticular-induced reflex suppression and, intracellularly, membrane potential hyperpolarization during active sleep. The following section presents a possible mechanism for this unique pattern of state-regulated response-reversal.

Supposition: Pontomesencephalic activation of the bulbar inhibitory area is effective only during active sleep due to the state-dependent activity of a coupling mechanism. When activated, the coupling mechanism opens a functional pathway to the bulbar region so that the discharge of pontomesencephalic neurons induces activity in bulbar neurons.

This aspect of the Model is based upon the supposition that activation of pontomesencephalic reticular sites leads to the discharge of bulbar reticular neurons, which then results in motoneuron hyperpolarization and reflex suppression during active sleep. This tonic driving of bulbar neurons is thought to be blocked during wakefulness.

In a summary of reticular homeostasis and the interplay between the reticular activating system and the bulbar inhibitory area, Paul Dell (1963) has stated that close functional interrelationships between these bulbar inhibitory mechanisms and the activating system have been demonstrated. "This suggests that each mesencephalic reticular activation brings into play a bulbar inhibitory mechanism. If one accepts some of the suggestions here made, then sleep would be only an extreme state resulting from an accumulation of bulbar inhibitory effects."

While there is no data to support state selectivity for this interaction, there are reports that pontomesencephalic neurons excite bulbar neurons (Ito, Udo, and Mano, 1970) by a process involving the induction of postsynaptic excitatory potentials (Mancia, Mariotti, and Spreafico, 1974). However, there is little other information at present that either supports or negates the proposed reticular-reticular interaction, and none that indicates state-selective driving of bulbar neurons by pontomesencephalic neurons.

DISCUSSION AND SUMMARY

The Model of Motor Control suggests a mechanism whereby a single reticular locus assumes the capability to express dual functions that are entirely state-dependent. This process is achieved by the blockade of a neuronal pathway during wakefulness from the pontomesencephalic reticular formation to the bulbar (medullary) inhibitory reticular formation, and the opening of this pathway during active sleep.

The basic postulate rests on the proposition that neuronal mechanisms operate selectively during active sleep to couple the pontomesencephalic reticular "activating" system (PMRAS) with the bulbar reticular "inhibitory" system (BRIS). The functional result of this coupling is the excitation of bulbar neurons by the discharge of pontomesencephalic neurons. The degree of bulbar discharge is thought to parallel, *pari passu*, the activity of pontomesencephalic neurons. Thus, as the activity of the PMRAS increases during active sleep (Steriade and Hobson, 1976), there is a corollary increase of BRIS discharge. A precise compensatory suppressive control is thereby generated by the BRIS which in turn functions to "contain" the arousing effects of PMRAS activation. By this means the animal is able to maintain the active sleep state with the minimum yet precisely appropriate level of "inhibitory" motor control necessary to compensate for the massed discharge of the PMRAS during active sleep. PMRAS activity presumably is triggered by internal generators that reside in the pons and mesencephalon, as well as by collaterals from almost all brain regions.

It is proposed that the level of arousal is therefore set by the activity of the PMRAS due to the discharge of internal generators or to driving by sensory, motor, and integrative systems. The coupling mechanism provides a means to convert a "general" excitatory drive during wakefulness to a "general" inhibitory drive during active sleep. It also explains, as described in a previous section of this chapter, how by PMRAS activation of the BRIS, all areas of the brainstem can induce motor suppression with a similar time course during active sleep.

The Model and the data serving as its foundation suggest that it is the pontomesencephalic reticular formation that controls the basic states of motor activity during wakefulness and active sleep. Since motor activity appears to be "a" or "the" key physiological process which provides the foundation for the behavioral expression of wakefulness and active sleep, the pontomesencephalic reticular formation may also play a crucial role in the initiation and/or maintenance of these states.

At this point in our studies we can go no further than to suggest a phenomenological hypothesis (i.e., the Model of Motor Control) that might be responsible for the conversion of an induced pattern of motor facilitation to one of suppression in a state-dependent fashion. We will test this hypothesis by various strategies, and if a selective coupling between reticular regions during active sleep is observed, then we should be able to determine if it is indeed the critical mechanism that underlies the phenomenon of response-reversal, and the extent of its participation in the control of sleep and waking states.

The Model as constructed is most likely sound, but its strength may be deceptive, for, as noted, some essential aspects have not been defined or accounted for when neither the literature nor our data bear directly upon them. As with any model, its chief value lies in providing a framework for organizing data as well as for generating concepts so that research strategies can be developed to determine their validity. We predict that the model proposed in this chapter will shortly be replaced by another as we define and then redefine our conceptual themes as new data is added to the store. However, the accumulated data which serve as the Model's foundation will remain, and we will continue to develop a more accurate, data-based picture of central neural function in the control of motor processes during sleep and wakefulness.

ACKNOWLEDGMENT

This research was supported by United States Public Health Service grant NS-09999.

REFERENCES

Babb, M. and M. H. Chase (1974). Reflex modulation during conditioned sensorimotor cortical activity, *Electroencephalogr. Clin. Neurophysiol.* **36**:357–365.

Babb, M. I., N. K. Wills-Lewis, and M. H. Chase (1976). Modification of masseteric reflex activity during sleep and wakefulness by stimulation of the ponto-medullary junction in cats, *Abstr. Annu. Meet. Soc. Neurosci.*, 6th, Toronto, p. 891.

Baldissera, F., G. Ettore, L. Infuso, M. Mancia, and C. A. Pagni (1966). Etude comparative des responses evoquées par stimulation des voies cortico-spinales pendant le sommeil et la veille, chez l'homme et chez l'animal, *Rev. Neurol. (Paris)* **115**:82–84.

Batini, C., C. Buisseret-Delma, and J. Corvisier (1976). Horseradish peroxidase localization of masticatory muscle motoneurons in cat, *J. Physiol. (Paris)* **72**:301–309.

Bremer, F. (1961). Neurophysiological mechanisms in cerebral arousal, pp. 30–56 in *The Nature of Sleep*, Wolstenholme, G. E. W. and M. O'Connor, eds. Little Brown and Co., Boston.

Burke, R. E. (1967). Composite nature of the monosynaptic excitatory post-synaptic potential, *J. Neurophysiol.* **30**:1114–1137.

Chandler, S. H., Y. Nakamura, and M. H. Chase (1977). Characterization of synaptic potentials induced by reticular formation stimulation on trigeminal motoneurons during sleep and wakefulness, *Abstr. Annu. Meet. Soc. Neurosci.*, 7th, Anaheim, p. 467.

Chase, M. H. (1974). Somatic reflex activity during sleep and wakefulness, pp. 249–267 in *Basic Sleep Mechanisms*, Petre-Quadens, O. and J. Schlag, eds. Academic Press, New York.

Chase, M. H. (1976). A model of central neural processes controlling motor behavior during active sleep and wakefulness, pp. 99–121 in *Mechanisms in Transmission for Signals for Conscious Behavior*, Desiraju, T., ed. Elsevier, Amsterdam.

Chase, M. H. and M. Babb (1973). Masseteric reflex response to reticular stimulation reverses during active sleep compared with wakefulness or quiet sleep, *Brain Res.* **59**:421–426.

Chase, M. H. and D. J. McGinty (1970a). Modulation of spontaneous and reflex activity of the jaw musculature by orbital cortical stimulation in the freely-moving cat, *Brain Res.* **19**:117–126.

Chase, M. H. and D. J. McGinty (1970b). Somatomotor inhibition and excitation by forebrain stimulation during sleep and wakefulness: orbital cortex, *Brain Res.* **19**:127–136.

Chase, M. H., D. J. McGinty, and M. B. Sterman (1968). Cyclic variations in the amplitude of a brain stem reflex during sleep and wakefulness, *Experientia (Basel)* **24**:47–48.

Chase, M. H., R. Monoson, K. Watanabe, and M. I. Babb (1976). Somatic reflex response-reversal of reticular origin, *Exp. Neurol.* **50**:561–567.

Chase, M. H. and N. Wills (1978). Brainstem control of masseteric reflex activity during sleep and wakefulness: medulla, *Exp. Neurol.*, in press.

Clemente, C. D., M. H. Chase, T. A. Knauss, E. K. Sauerland, and M. B. Sterman (1966). Inhibition of a monosynaptic reflex by electrical stimulation of the basal forebrain or the orbital gyrus in the cat, *Experientia (Basel)* **22**:844–848.

Dell, P. (1963). Reticular homeostasis and critical reactivity, pp. 82–103 in *Brain Mechanisms (Prog. Brain Res.*, Vol. 1), Moruzzi, G., A. Fessard, and H. H. Jasper, eds. Elsevier, Amsterdam.

Dell, P., M. Bonvallet, and A. Hugelin (1961). Mechanisms of reticular deactivation, pp. 86–107 in *The Nature of Sleep*, Wolstenholme, G. E. W. and M. O'Connor, eds. Little Brown and Co., Boston.

Enomoto, T., K. Fukuoka, Y. Imai, M. Kako, Y. Kaneko, M. Mishimagi, A. Ono, and K. Kubota (1968). Masseteric monosynaptic reflex in chronic cat, *Jpn. J. Physiol.* **18**:169–178.

Goldberg, L. J. (1968). Electrophysiological studies of oral-pharyngeal reflex mechanisms. Ph.D. Dissertation, Department of Anatomy, University of California, Los Angeles.

Granit, R. (1970). *The Basis of Motor Control.* Academic Press, New York.

Granit, R., J. O. Kellerth, and T. D. Williams (1964*a*). Intracellular aspects of stimulating motoneurones by muscle stretch, *J. Physiol. (London)* **174**:435–452.

Granit, R., J. O. Kellerth, and T. D. Williams (1964*b*). "Adjacent" and "remote" postsynaptic inhibition in motoneurones stimulated by muscle stretch, *J. Physiol. (London)* **174**:453–472.

Hodes, R. and J. K. Suzuki (1965). Comparative thresholds of cortex, vestibular system and reticular formation in wakefulness, sleep and rapid eye movement periods, *Electroencephalogr. Clin. Neurophysiol.* **18**:239–248.

Hugelin, A. (1961). Integrations motrices et vigilance chez l'encephale isolé. II. Controle reticulaire des voies finales communes d'ouverture et de fermeture de la gueule, *Arch. Ital. Biol.* **99**:244–269.

Hugelin, A., and M. Bonvallet (1957). Tonus cortical et controle de la facilitation motrice d'origine reticulaire, *J. Physiol. (Paris)* **49**:1171–1200.

Ito, M., M. Udo, and N. Mano (1970). Long inhibitory and excitatory pathways converging onto cat reticular and Deiters' neurons and their relevance to reticulofugal axons, *J. Neurophysiol.* **33**:210–226.

Iwama, K. and T. Kawamoto (1966). Responsiveness of cat motor cortex to electrical stimulation in wakefulness, pp. 54–63 in *Correlative Neurosciences, Part B: Clinical Studies (Prog. Brain Res.*, Vol. 21B), Tokizane, T. and J. P. Schade, eds. Elsevier, Amsterdam.

Jouvet, M. (1972). The role of monoamines and acetylcholine-containing neurons in the regulation of the sleep-waking cycle, *Ergeb. Physiol.* **64**: 166–307.

Jouvet, M., F. Michel, and J. Courjon (1960). Etude E.E.G. du sommeil physiologique chez le chat intact, décortiqué et mesencephalique chronique, *Rev. Neurol. (Paris)* **102**:309–310.

Kaada, B. R. (1951). Somato-motor, autonomic and electroencephalographic responses to electrical stimulation of "rhinencephalic" and other structures in primates, cat and dog, *Acta Physiol. Scand.* **24** (Suppl. 83):1–285.

Kidokoro, Y., K. Kubota, S. Shuto, and R. Sumino (1968). Reflex organization of cat masticatory muscles, *J. Neurophysiol.* **31**:695–708.

Kuno, M. (1971). Quantum aspects of central and ganglionic synaptic transmission in vertebrates, *Physiol. Rev.* **51**:647–678.

Landgren, S. and K. A. Olsson (1976). Localization of evoked potentials in the digastric, masseteric, supra- and intertrigeminal subnuclei of the cat, *Exp. Brain Res.* **26**:299–318.

Mancia, M., M. Mariotti, and R. Spreafico (1974). Caudo-rostral brain stem reciprocal influences in the cat, *Brain Res.* **80**:41–51.

McIntyre, A. K. (1951). Afferent limb of the myotatic reflex arc, *Nature (London)* **168**:168–169.

Mizuno, N., A. Konishi, and M. Sato (1975). Localization of masticatory motoneurons in the cat and rat by means of retrograde axonal transport of horseradish peroxidase, *J. Comp. Neurol.* **164**:105–116.

Mizuno, N., E. R. Sauerland, and C. D. Clemente (1969). Projections from the orbital gyrus in the cat. I: To brain stem structures, *J. Comp. Neurol.* **133**: 463–476.

Moruzzi, G. (1972). The sleep-waking cycle, *Ergeb. Physiol.* **64**:1–165.

Moruzzi, G. and H. W. Magoun (1949). Brain stem reticular formation and activation of the EEG, *Electroencephalogr. Clin. Neurophysiol.* **1**:455–473.

Nakamura, Y. (1978). Brain stem neuronal mechanisms controlling the trigeminal motoneuron activity, in press.

Nakamura, Y., L. J. Goldberg, S. H. Chandler, and M. H. Chase (1978). Intracellular analysis of trigeminal motoneuron activity during sleep in the cat, *Science* **199**:204–207.

Nakamura, Y., L. J. Goldberg, and C. D. Clemente (1967). Nature of suppression of the masseteric monosynaptic reflex induced by stimulation of the orbital gyrus of the cat, *Brain Res.* **6**:184–198.

Pompeiano, O. (1967a). Sensory inhibition during motor activity in sleep, pp. 323–375 in *Neurophysiological Basis of Normal and Abnormal Motor Activities*, Yahr, M. D. and D. P. Purpura, eds. Raven Press, New York.

Pompeiano, O. (1967b). The neurophysiological mechanisms of the postural and motor events during desynchronized sleep, pp. 351–423 in *Sleep and Altered States of Consciousness (Res. Publ. Assoc. Res. Nerv. Ment. Dis.*, Vol. 45), Kety, S. S., E. V. Evarts, and H. L. Williams, eds. Williams and Wilkins, Baltimore.

Rossi, G. F. and A. Zanchetti (1957). The brain stem reticular formation: anatomy and physiology, *Arch. Ital. Biol.* **95**:199–435.

Sauerland, E. K., Y. Nakamura, and C. D. Clemente (1967). The role of the lower brain stem in cortically induced inhibition of somatic reflexes in the cat, *Brain Res.* **6**:164–180.

Steriade, M. and J. A. Hobson (1976). Neuronal activity during the sleep-waking cycle, *Prog. Neurobiol.* **6**:155–376.

Szentagothai, J. (1948). Anatomical considerations of monosynaptic reflex arcs, *J. Neurophysiol.* **11**:445–454.

Tarchanoff, J. (1894). Quelques observations sur le sommeil normal, *Arch. Ital. Biol.* **21**:318–321.

Thelander, H. E. (1924). The course and distribution of the radix mesencephalica trigemini in the cat, *J. Comp. Neurol.* **37**:207–220.

Wills, N. and M. H. Chase (1978). Brainstem control of masseteric reflex activity during sleep and wakefulness: mesencephalon and pons, *Exp. Neurol.*, in press.

Wills-Lewis, N. K. and M. H. Chase (1976). Mesodiencephalic influences on reflex activity during states of sleep and wakefulness, p. 35 in *Sleep Research*, Vol. 5, Chase, M.H., M. M. Mitler, and P. L. Walter, eds. Brain Information Service/Brain Research Institute, University of California, Los Angeles.

SEROTONERGIC NEURONAL ACTIVITY AND AROUSAL OF FEEDING IN *APLYSIA CALIFORNICA*

Klaudiusz R. Weiss and Irving Kupfermann

Columbia University College of Physicians and Surgeons, and New York State Psychiatric Institute, New York City

INTRODUCTION

In an attempt to deal with the great variety of plasticity that behavior exhibits, psychologists have invoked two classes of explanatory variables—learning and motivational states. The distinction between these two types of variables is not always clear-cut, but common use of the terms generally follows certain regularities. Behavioral changes attributed to learning are typically long-lasting and are the result of previous patterns of external sensory information. Behavioral changes attributed to alterations of motivational state are relatively short-lasting, frequently occur cyclically and spontaneously, and are the result of changes in the bodily state of the organism. Thus, for example, an organism that initially will accept food may alter its behavior and reject food, either because it has *learned* that the food is associated with a noxious stimulus, or because its *motivational state* has altered, that is, it is no longer hungry. Certain states, such as arousal, are difficult to categorize. Arousal is usually considered to be a motivational variable, since arousal states are relatively short-lived and can occur spontaneously and cyclically. However, as in learning, arousal states can also be induced by specific patterns of exteroceptive stimuli. One problem with understanding the fundamental relationship of intervening variables such as learning and arousal is that they typically are not directly studied; their properties are inferred on the basis of behavioral or crude neurophysiological indices. Clarification of the nature of motivational states and their relationship to learning will probably come only after direct study of the neural basis of these hypothetical entities.

We have been analyzing the neural basis of feeding behavior in *Aplysia* in an attempt to develop a model system for the study of motivational or state variables. *Aplysia* was chosen for study because, in addition to the well-known advantages of its nervous system for cellular study, its feeding behavior exhibits abundant plasticity and appears to be under the control of a variety of motivational variables which in many respects are behaviorally similar to those seen in higher animals. Thus, for example, a large meal produces a state of satiation in which the animal refuses food and becomes inactive. In this paper, we will consider a second major class of state variables shown by *Aplysia:* food arousal.

The food-arousal state in *Aplysia* can be induced by exposing a quiescent animal to a seaweed stimulus in the water (Frings and Frings, 1965; Kupfermann, 1974; Preston and Lee, 1973). Within 10 to 60 sec, the formerly quiescent animal will begin to orient and assume a characteristic feeding posture (Figure 1A) in which it anchors its tail to the substrate and waves its neck to and fro, presumably to localize the stimulus. If the food is not directly contacted, animals will frequently begin to locomote. When food makes contact with the lips, a stereotyped biting response is elicited. This response consists of protraction of the radula, grasping of the seaweed between the radula halves, and finally, retraction of the radula and seaweed into the buccal cavity.

The effect of the food-arousal state on the biting response can be quantified by measuring two variables: (1) the strength of the biting response, and (2) the time between successive biting responses elicited by continuous exposure to food. Following initial exposure to food there is a progressive build-up in the strength of biting (Figure 1C), as well as a decrease in the time between successive rhythmic biting movements (Figure 1B). A decrease of biting latency following exposure to food persists for at least half an hour after the food is removed. As in higher animals, exposure of *Aplysia* to food not only modifies feeding responses, but also alters other reflexes not directly related to feeding. For example, food arousal modifies defensive withdrawal reflexes of the head and mantle organs (Advokat, Carew, and Kandel, 1976; Preston and Lee, 1973). In addition, during the build-up of the food-arousal state as measured by changes in interresponse times of biting, there is a parallel (Figure 1D) build-up in the heart rate of the animals (Dieringer, Koester, and Weiss, 1978).

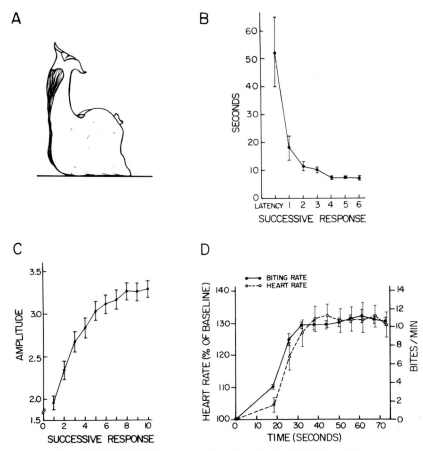

FIGURE 1. Behavioral features of food arousal in *Aplysia*. A: *Aplysia* in its characteristic feeding posture during the arousal state elicited by brief exposure to food (from Kupfermann, 1974). B: Decrease of the successive interbite intervals of an animal continuously stimulated with food (from Susswein, Weiss, and Kupfermann, 1978). C: Enhancement of the strength of successive biting responses of an animal continuously stimulated with food. The strength of the biting response was measured on an arbitrary 1 to 4 scale (Susswein et al., 1978). D: Parallel increase of the speed of biting and of the heart rate of an animal continuously exposed to food (Dieringer et al., 1978).

Some of the various behavioral characteristics of food arousal in *Aplysia* have been summarized in Table 1. Comparison with arousal states in vertebrates reveals surprising correspondence. First, as in higher animals, arousal in *Aplysia* produces changes in spontaneous

TABLE 1. *Behavioral concomitants of food arousal*[1]

Spontaneous activity		Reflex activity	
Feeding behavior	Non-feeding	Feeding behavior	Non-feeding
head waving (↑)	heart rate (↑) siphon tone (↓)	biting latency (↓) biting strength (↑) biting frequency (↑)	head withdrawal (↓) siphon withdrawal (↓) inking (↓)

[1] Arrows indicate whether arousal is associated with an increase (↑) or decrease (↓) of the incidence, strength, or speed of the response. For references, see text.

behavior as well as alterations of reflexive responses. Second, arousal produces both specific and generalized changes in behavior. In our example, specific changes are those related rather directly to feeding, such as head waving and biting. Generalized changes are those not directly related to feeding, such as modification of heart rate and siphon reflexes. Finally, arousal can potentiate certain responses and suppress or inhibit others. The net result of all of the behavioral changes associated with arousal is to increase the efficiency of the animal in dealing with a certain type of environment. Behavioral analysis of food arousal in *Aplysia* indicates that this state affects virtually every behavior in the animal, but not surprisingly, it has the most pronounced effects on feeding behavior. In this paper we will concentrate on food-arousal effects on the biting response, specifically on the progressive increase in the strength of biting and decrease of interbite intervals following exposure of the animal to food. We will emphasize the possible role of a pair of giant serotonergic cells found in the cerebral ganglia. These cells, termed the metacerebral cells (MCCs) (Kandel and Tauc, 1966), were of special interest for a number of reasons.

First, homologous cells appear to be present in most, if not all, of several thousands of varied species of pulmonate and opisthobranch molluscs (Berry and Pentreath, 1976; Kandel and Tauc, 1966; Nabias, 1894; Senseman and Gelperin, 1974; Weiss and Kupfermann, 1976). Second, the transmitter agent utilized by the cells had been extensively studied and had been identified as serotonin (Eisenstadt, Goldman, Kandel, Koike, Koester, and Schwartz, 1973; Weinreich, McCaman, McCaman, and Vaughn, 1973; Paupardin-Tritsch and Gerschenfeld, 1973). Finally, Cottrell and co-workers (Cottrell, 1970; Cottrell and Macon, 1974) showed that the MCCs of pulmonate molluscs make synaptic connections to cells in the buccal ganglion, which contains motor neurons for the muscles that execute biting and swallowing movements. Paupardin-Tritsch and Gerschenfeld (1973) similarly showed that the MCCs of *Aplysia* make synaptic connections to a number of unidentified cells in the buccal ganglion.

In this paper, we will present evidence that activity of the MCC together with the plastic properties of the terminals of buccal motor neurons may account for several aspects of the behavioral changes that occur in the biting response during the food-arousal state.

RESULTS

Chronic Recording

The metacerebral cells of *Aplysia* are located in the cerebral ganglion. By recording extracellularly from various peripheral nerves and firing the MCC by intracellular current, we showed that the cell has two main axonal processes of large diameter (Figure 2A). One axon enters the posterior lip nerve, which innervates the lip and possibly the anterior regions of the buccal mass. The other main process of the MCC enters the cerebral-buccal connective. This process enters the buccal ganglion and ramifies into numerous branches that enter all of the main buccal nerves that innervate the muscles of the buccal mass. The axon distribution of the MCC clearly suggested some role of this cell in feeding behavior.

FIGURE 2. Distribution of the main axons of the MCC and MCC activity during feeding. A1: Semidiagrammatic distribution of axons of the left MCC in cerebral and buccal ganglia nerves. The distribution of axons of the right MCC is a mirror image. A2: Recordings of extracellular spikes in the nerves, produced by intracellular stimulation of the MCC. A2a: Recordings in buccal nerves. A2b: Recordings in cerebral nerve and connectives. Abbreviations: B.n., buccal nerve; OES. n., oesophageal nerve; C-B conn., cerebral-buccal connective; RAD. n., radula nerve; a. LIP n., m. LIP n., and p. LIP n., anterior, medial, and posterior lip nerve, respectively; P.-P. conn., pleuro-pedal-cerebral connective (Weiss and Kupfermann, 1976). B: Chronic recordings from a freely moving animal. B1a: Simultaneous records from the posterior lip nerve and the cerebral-buccal connective. At the dark arrow the rhinophores were touched with food. At the open arrow the lips were touched with food. Food was kept in contact with the animal from the first arrow until the end of the record. B1b: Faster sweep of a part of record shown in B1a, to illustrate the wave form and synchrony of spikes in lip nerve and connective. B2: Simultaneous intracellular record of MCC spikes elicited by intracellular stimulation and extracellular recording from the lip nerve and connective. These data were obtained from the isolated nervous system of the animal whose records are shown in part A above, utilizing the same extracellular electrodes (Weiss et al., 1978).

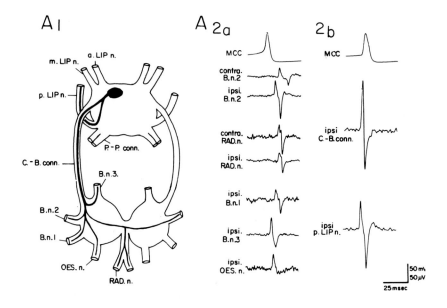

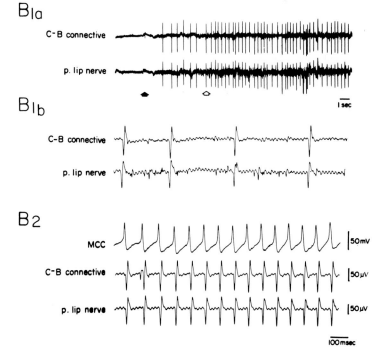

The large diameter of the axons of the MCC in both the lip nerve and cerebral-buccal connective made possible the recording of its activity in free-moving animals. For these experiments a special chronic suction electrode was fabricated. Animals were immobilized with magnesium, and miniature suction electrodes were attached to the cut end of one or both of the nerve trunks containing a MCC axon. The following day it was possible to record MCC activity (Figure 2B) during normal feeding behavior. In selected cases, animals were sacrificed and the identity of the extracellular spike was established by impaling the MCC and comparing its extracellularly recorded spike with that obtained in the free-moving animal. The results of these experiments clearly indicated a relationship between MCC activity and feeding behavior. In quiescent animals the MCC is usually silent; 5 to 60 sec after food contacts the lips or tentacles of the animal, the MCC begins to fire at a rate of 1 to 10/sec. The onset of MCC activity closely corresponds to the first occurrence of behavioral signs of food arousal, such as head waving. If the animals are now repeatedly fed pieces of seaweed, the mean spike frequency of the MCC decreases throughout the meal, roughly in parallel with the decrease of response speed and response strength seen as animals satiate. At the point at which animals will no longer respond to food, food also becomes ineffective in firing the MCC (Kupfermann and Weiss, unpublished observations).

Central Effects of the MCC

Chronic recording provided evidence that activity of the MCCs is intimately associated with feeding behavior. To provide more direct evidence for a role of the MCCs in feeding, a preparation of the isolated nervous system was used in which it was possible to impale the MCC while recording extracellularly from the nerves innervating the buccal muscles. Bursts of cyclical activity in the buccal nerves are associated with organized movements of the buccal mass and probably represent some aspect of the biting or swallowing central program. In preparations in which spontaneous burst activity of the buccal nerves was not occurring, firing of a MCC did not trigger cyclical bursts. On occasion, high-frequency firing of the MCC could elicit a single burst of activity from a buccal nerve in a quiescent preparation, but not even this was seen at physiological rates of MCC activity. In contrast, if spontaneous burst output was present in buccal nerves, firing of a MCC could increase the frequency of occurrence of the bursts (Figure 3). Thus, these experiments suggested that firing of the MCC could modulate ongoing activity of the buccal ganglion but was minimally effective in initiating activity. Furthermore, the data suggested that the progressive increase of frequency of biting responses at the onset of a meal may be related to the fact that the MCC becomes active at the onset of a meal. If activity of the MCC accounts for increased biting frequency during arousal, what accounts for the progressive build-up in response strength during increased arousal? One factor that we feel contributes to increased response strength at the onset of a meal is post-tetanic potentiation at the motor-neuron-to-muscle synapse.

Post-Tetanic Potentiation

To study the properties of the neuromuscular system of the buccal mass, one experimentally advantageous muscle, the accessory radula closer muscle (ARC muscle), was selected for intensive investigation. The properties of this muscle appear to be representative of other buccal muscles of *Aplysia* (Cohen, Weiss, and Kupfermann, 1974, 1978; Orkand and Orkand, 1975). The ARC is innervated by three or four motor neurons, two of which can be reliably located in the buccal ganglion. Each motor neuron produces end-plate junction potentials (EJPs) in each fiber of the muscle. Contraction occurs when the EJPs depolarize the muscle beyond approximately -30 mV. The muscle fibers never exhibit an action potential, and the force of contraction is a monotonic function of the amount and duration of depolarization. For this reason, any changes in the size of the EJP are directly translated into changes of the force of contraction. (For a different type of organization of gastropod buccal muscle, see the work of Kater, Heyer, and Hegmann, 1971, on the snail.)

When the motor neurons are fired in bursts similar to those that occur during rhythmic burst output of the buccal ganglion (Figure 4B), the EJPs exhibit facilitation within a burst and post-tetanic potentiation between bursts, even when the bursts are separated by 10 sec or more. The amount of post-tetanic potentiation is greater with shorter interburst intervals. Thus, for example, the EJPs are larger when motor neuron bursts are given every 6 sec than when given every 10 sec. As expected, repeated bursts of a motor neuron result in a progressive increase of the force of contraction, and the contractions grow more rapidly with 6/sec motor neuron bursts than with 10/sec bursts (Figure 4A). It thus seems likely that part of the reason for a progressive increase in the force of contractions at the onset of a meal is that the EJPs

BUCCAL
NERVE 2
MCC

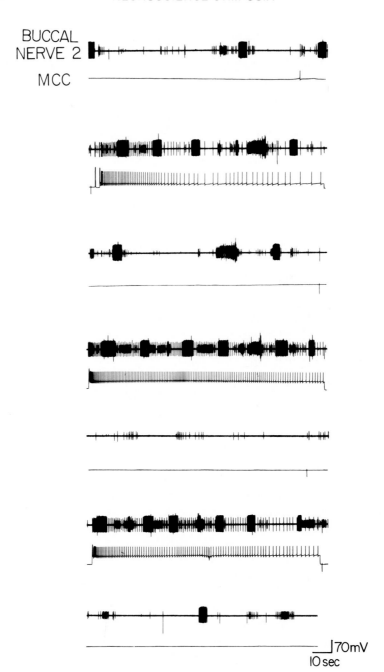

70mV
10 sec

progressively increase. This increase may be the outcome of two factors: (1) simple post-tetanic potentiation, and (2) an enhancement of the potentiation due to the action of the MCC, which shortens the interburst interval and thus decreases the amount of decay of potentiation between successive bursts.

An additional factor that might contribute to enhancement of contraction is a direct excitatory effect of the MCC on certain motor neurons. The excitatory effect of the MCC is rarely sufficient to fire the motor cells, but could enhance the frequency of their firing when the motor cells are provided with other sources of excitation. Enhancement of the frequency of a motor neuron burst would result in greater muscle contraction because of more effective temporal summation of the EJP as well as enhanced facilitation.

Peripheral Effect of the MCC

The central effects of the MCC could contribute both to the enhancement of contraction and to increased biting frequency during arousal of feeding. What then is the function of the peripheral branches of the MCC that enter all of the nerves innervating the buccal muscles? Our data indicate that these peripheral branches may provide an additional means of enhancing muscle contraction. As appears to be the case for its central effects, the effect of the MCC on the peripheral muscle is purely modulatory. Thus, in itself, firing of the MCC produces neither contraction nor relaxation of buccal muscles. However, activity of the MCC produces an increase in the force of contraction produced by a fixed burst of spikes in a buccal muscle motor neuron (Figure 5). The enhancement of buccal muscle contraction is proportional to the number of MCC spikes, and clear enhancement occurs at the relatively low rates of MCC firing that are seen in recordings from free-moving animals during feeding. The enhancement of muscle contraction produced by MCC stimulation can magnify the effects of post-tetanic potentiation of motor neuron terminals. Thus, when repeated bursts of motor neuron

---

FIGURE 3. Effect of MCC activity on spontaneous cyclic burst activity of the buccal ganglion. Fourteen min of continuous recording are shown. Each pair of traces represents 2 min of simultaneous recording from the MCC and buccal nerve 2. During alternate 2-min periods, by means of hyperpolarizing or depolarizing current, the MCC was either silenced or was made to fire at .5–2 spikes per sec. During periods when the MCC was fired, burst activity clearly increased. One of the large spikes seen in nerve 2 is the extracellular record of an axon of the MCC and corresponds one for one with the intracellular spikes of the MCC. From Weiss et al. (1978).

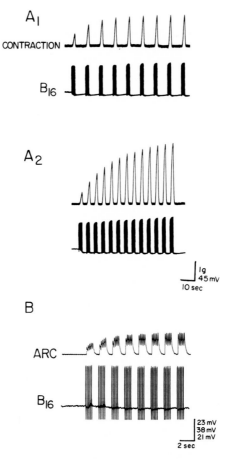

FIGURE 4. A: Post-tetanic potentiation and build-up of contraction. B16 was stimulated in bursts at 23 Hz for 2 sec while muscle tension was monitored with a strain gauge. Interburst interval was 10 sec (A1) or 6 sec (A2). With the shorter interburst interval, the contraction showed more build-up (Cohen et al., 1978). B: Facilitation and post-tetanic potentiation of EJPs from motor neuron B16.

spikes are given together with MCC stimulation, the rate and final level of muscle contraction is enhanced over that seen when no MCC spikes are given (Weiss, Cohen, and Kupfermann, 1975, 1978).

Mechanism of MCC Enhancement of Buccal Muscle Contraction

In order to insure that the increase of muscle contractions following MCC stimulation was due to the action of the MCC at the peripheral

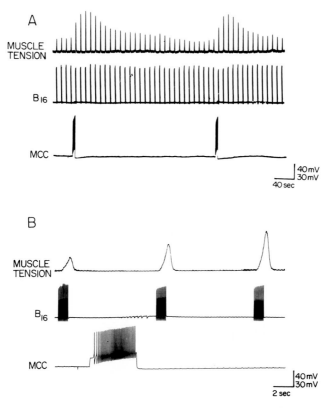

FIGURE 5. Metacerebral cell potentiation of ARC muscle contraction. A: Contractions of the ARC muscle were produced every 10 sec by firing a brief burst of spikes in the ARC motor neuron B16. In order to control the number and frequency of spikes, every spike was triggered by an individual depolarizing pulse. This stimulation procedure resulted in reliable and reproducible contractions of the ARC muscle. Immediately following a brief burst of MCC spikes there was an increase of the contraction elicited by motor neuron stimulation, and subsequent contractions continued to increase for 30 to 40 sec. Contraction size then gradually returned to control over a period of 1 to 2 min. B: Expanded version of the second MCC stimulation period shown in A. From Weiss et al. (1978).

neuromuscular system rather than at the central ganglionic level, two types of experiments were performed. First, we blocked all central chemical synaptic connections of the MCC by bathing the ganglion, but not the muscle, in a solution of high Mg^{2+} low Ca^{2+}. Under these conditions, the MCC continued to potentiate muscle contractions. Second, we applied low concentrations of the transmitter (serotonin,

10^{-8} to 10^{-9} M) of MCC to the muscle, and this simulated the effects of MCC stimulation, i.e., muscle contractions were potentiated.

The peripheral effects of the MCC suggested that this neuron might be exerting an unusual action at buccal muscle. To investigate this mechanism, the strength of contraction was measured simultaneously with intracellular recordings from the MCC, motor neuron, and muscle fibers. In the course of many hundreds of penetrations of muscle fibers, firing of the MCC was never observed to produce any shift in the resting potential of the fibers. Furthermore, as judged by the time course of decay of the EJP, the MCC did not alter passive membrane conductance. On the other hand, firing of the MCC produced a small but distinct enhancement of the size of the EJP produced by motor neuron stimulation (Figure 6). For motor neuron B15, the time course of the enhancement of EJP size paralleled the potentiation of muscle contraction. Thus, both the peak for enhancement of EJP size and the peak potentiation of muscle contraction occurred approximately 30 sec after cessation of MCC firing. These observations suggested that a form of heterosynaptic facilitation could account for the MCC effects. However, observation of the effects of MCC stimulation on the EJP produced by motor neuron B16 revealed a very different pattern (Figure 7). For this motor neuron, peak enhancement of the EJP occurred immediately after firing of the MCC. Thirty sec later, when the muscle contraction was maximal, the EJP size had almost returned to the control level. These results suggested that for motor neuron B15, MCC enhancement of contraction may involve some direct effect on excitation-contraction coupling. Even for neuron B16, the enhancement of EJP size following MCC stimulation appeared to be too small to fully explain the observed increase of the force of muscle contraction. Consequently, it seemed possible that the MCC effects of B16 could also involve a direct effect on excitation-contraction coupling. To test this idea, we attempted to determine whether the MCC could enhance contractions produced by B16 even when the EJP size did not increase (Figure 8A,B). To prevent the usual increase of EJP size following MCC stimulation, the interval between motor neuron bursts was increased following the MCC stimulation. Since the EJPs exhibit substantial post-tetanic potentiation, an increase in the time between motor neuron bursts permits the post-tetanic potentiation to decay, resulting in a considerable decrease of EJP magnitude. This decrease of the EJP ordinarily leads to a considerable reduction in the force of contraction. When the EJP size was reduced in this manner, the MCC nevertheless

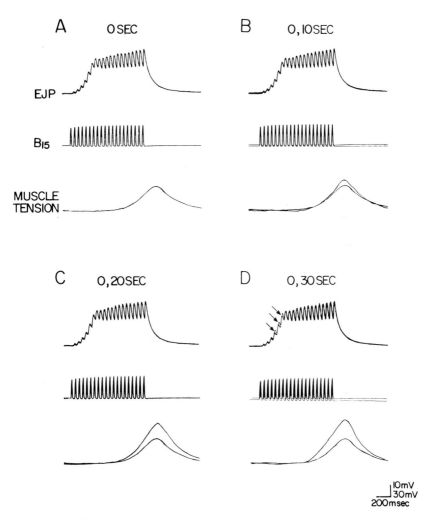

FIGURE 6.   Effect of MCC stimulation on muscle tension and size of EJP produced by motor neuron B15. A: Control recording before MCC was fired. B, C, and D: Recording 10, 20, and 30 sec after the MCC was fired. For comparison, each of these traces has been superimposed on the control recording. From Weiss et al. (1978).

enhanced the force of contraction. In other words, this experiment suggested that the MCC could enhance muscle contraction despite an actual reduction in EJP size.

Further evidence consistent with the hypothesis that the MCC acts

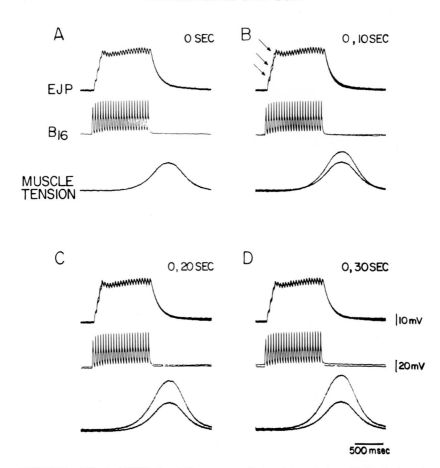

FIGURE 7.   Effect of MCC stimulation on muscle tension and size of EJP produced by motor neuron B16. A: Control recording before MCC was fired. B, C, and D: Recording 10, 20, and 30 sec after the MCC was fired. For comparison, each of these traces has been superimposed on the control recording. From Weiss et al. (1978).

directly on the muscle was obtained in experiments (Figure 8C) showing that stimulation of the MCC could enhance contractions produced by applications of brief pulses of acetylcholine, the excitatory transmitter at the muscle (Cohen et al., 1978). Under these conditions, potentiation of contraction obviously could not be a consequence of a presynaptic action at the motor neuron.

Role of cAMP in Potentiation of Muscle Contraction

For a number of reasons we suspected that potentiation of muscle contraction might be mediated by a second messenger such as cAMP: (1) The onset and decay of the effects of the MCC are remarkably slow for conventional synaptic transmission; (2) the action of the MCC on buccal muscle appeared to involve an unusual neurotransmitter mechanism in which there is no alteration of resting membrane potential or conductance; (3) serotonin and nerve stimulation enhance cAMP synthesis in the abdominal ganglion of *Aplysia* (Cedar, Kandel, and Schwartz, 1972; Cedar and Schwartz, 1972); (4) cAMP has been implicated in mediating response enhancement of gill-withdrawal reflexes in *Aplysia* (Brunelli, Castellucci, and Kandel, 1976; Kandel, Brunelli, Byrne, and Castellucci, 1976) and increased force of vertebrate cardiac muscle following adrenergic agonists.

In initial experiments (Weiss, Schonberg, Cohen, Mandelbaum, and Kupfermann, 1976), we investigated whether serotonin could affect cAMP synthesis in buccal muscle. As indicated earlier, application of low doses of serotonin to the muscle mimics the potentiation that the MCC can produce. In a cell-fragment preparation of buccal muscle, serotonin produced a dose-dependent enhancement of the synthesis of cAMP, indicating that the muscle contains a serotonin-sensitive adenyl cyclase. Application of serotonin to the intact muscle produced an exceptionally powerful stimulation of cAMP synthesis (Figure 9A). At 10^{-4} M a maximum 200-fold increase of cAMP synthesis was seen. The threshold for stimulation was approximately 10^{-8} M, which is comparable to the threshold of serotonin concentration needed to produce enhancement of muscle contraction when added to the bath. Control experiments with phosphodiesterase indicated that the product of synthesis is bona fide cAMP. In addition, we found that as well as enhancing synthesis of cAMP, serotonin increased the total tissue levels of cAMP (Figure 9A) as measured by the isotopic displacement method of Gillman (1970).

The experimental accessibility of the MCC afforded a unique opportunity to determine whether activity that was limited to this individual neuron could release sufficient transmitter to alter the rate of synthesis of cAMP. In five experiments, the MCC innervating the ipsilateral buccal muscle was fired for about 40 sec. The symmetrical buccal muscle, not innervated by the contralateral cell, served as a

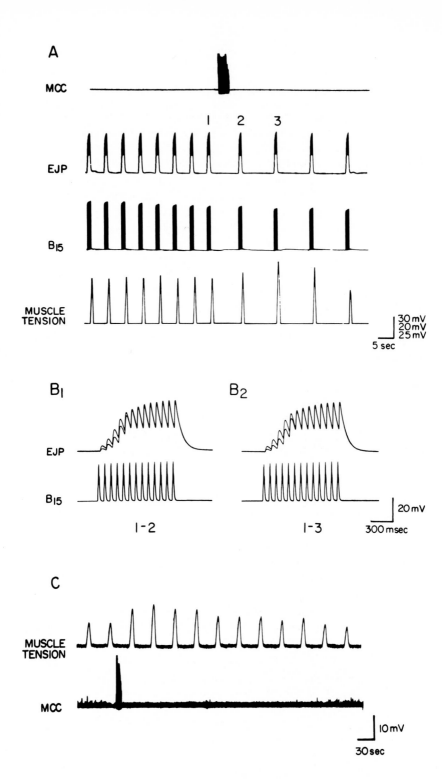

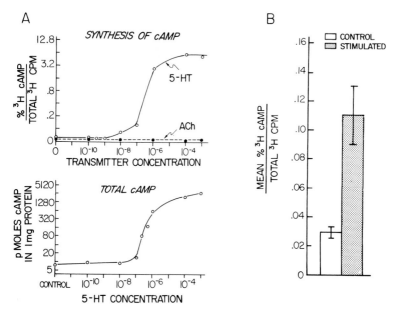

FIGURE 9.   A: Dose-response for synthesis and total cAMP level in ARC muscles exposed to serotonin. To measure synthesis, the muscle was incubated in [³H]adenosine in order to obtain a pool of labeled ATP, which served as a substrate for cAMP synthesis. B: Effect of stimulating individual metacerebral cells on the rate of cAMP synthesis in the ARC muscle. (Five experimental and five matched, nonstimulated controls. Standard error shown.)

control. In all experiments, the muscle ipsilateral to the MCC that was fired had a higher level of synthesis than the control (Figure 9B). The mean difference between the two muscles was 3.5-fold (p < .05, two-tailed *t* test). Additional experiments replicated these findings and showed that the product was bona fide cAMP. In a last set of experiments, we examined whether cAMP or cAMP analogues could

FIGURE 8.   Evidence that MCC affects excitation-contraction coupling. A: Effect of MCC on muscle contraction when the size of the EJP is experimentally reduced by increasing the interburst interval of the motor neuron. Slow record showing the paradigm used. B1: Superimposition of the control EJP (1) and the first EJP (2) that followed the simultaneous stimulation of the MCC and the increase of interburst interval. B2: The superimposition of control trace (1) and the second burst (3) after MCC stimulation. C: Effect of MCC stimulation on muscle contractions produced by direct application of acetylcholine to the muscle. The muscle was placed in a small flow-through perfusion chamber. Seawater was continuously perfused at 2 ml/sec and discrete amounts of ACh were added to the inflow at constant intervals. Peak ACh concentration was approximately 10⁻⁵ M. From Weiss et al. (1978).

mimic the effects of the MCC on muscle contraction. Bath application of cAMP did not enhance muscle contraction. Since cAMP penetrates cells poorly and is rapidly destroyed by phosphodiesterase present in buccal muscle, we examined the effects of the phosphodiesterase-resistant analogues 8-PCPT-cAMP and 8-BT-cAMP. Both of these agents produced an initial depression of muscle contraction followed by a dramatic and long-sustained potentiation of contraction (Figure 10). Intracellular

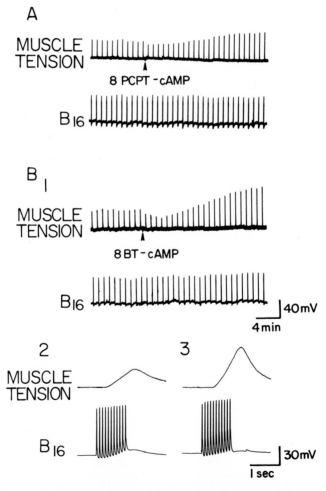

FIGURE 10. Effect of 8-Benzylthio-cAMP (8-BT-cAMP) and 8-Parachlorophenylthio-cAMP (8-PCPT-cAMP) on the strength of contractions elicited by stimulating the ARC motor neuron. B: Expanded record of A2 showing that the increase of muscle contractions was not due to changes in the number of motor neuron spikes.

recordings from the muscle during application of cAMP analogues revealed that EJP size was not enhanced. On the contrary, for reasons we do not understand, the analogues produced an immediate and prolonged depression of EJP amplitude. The enhancement of contraction occurred despite the reduction in EJP amplitude (Figure 11), suggesting that the cAMP analogues were acting directly on the muscle to enhance excitation-contraction coupling. These findings, together with the previous results showing that the MCC can enhance cAMP synthesis, provide support for the hypothesis that the direct effect of the MCC on excitation-contraction coupling may utilize cAMP as a second messenger.

DISCUSSION

The present results suggest that arousal of feeding behavior in *Aplysia* is a remarkably complex process. Although the data are incomplete, the indication is that at least five classes of events at different loci underlie arousal. Four of the effects we describe are due to the action of the MCC. Data from several other species of gastropod molluscs support the hypothesis of a role of the MCC in mediating arousal of biting and grasping behavior (Berry and Pentreath, 1976; Gelperin and Chang, 1976; Gillette and Davis, 1977; Granzow and Kater, 1978). Central actions of the MCC are exerted at the level of buccal motor neurons and

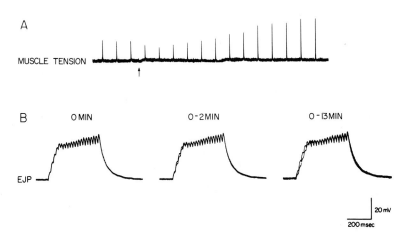

FIGURE 11. Effects of 8-BT-cAMP on muscle contraction and the size of EJPs. A: Enhancement of muscle contractions evoked by stimulating motor neuron B16. B: Records of EJPs prior to the application of 8-BT-cAMP, and superimpositions of the control EJPs on EJPs recorded 2 min and 15 min after the application of 8-BT-cAMP. The EJPs following 8-BT-cAMP are depressed, but the contraction is enhanced.

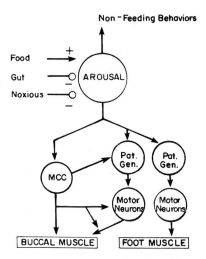

FIGURE 12. Hypothetical model of the organization of the food-arousal system in *Aplysia*. A central arousal system is postulated to provide synaptic input to neural systems controlling feeding behavior or behaviors not directly related to feeding. The arousal system is excited by food stimuli and is inhibited by noxious stimuli to the skin and stretch of the gut. Many of the effects of the arousal system on the biting response are executed by means of the activity of the MCC. In addition to affecting buccal muscles, EJPs, and motor neurons, the MCC may affect a postulated pattern-generating system that drives the motor neurons of the buccal mass.

at a hypothesized pattern generator. Peripheral actions of the MCC appear to operate by affecting excitation-contraction coupling directly and by enhancing the size of EJPs by a presynaptic or postsynaptic mechanism. Figure 12 summarizes our current thinking about the organization of the food-arousal system in *Aplysia.*

In addition to the action of the MCC, one aspect of food arousal, i.e., build-up of response strength, may arise, in part, because of post-tetanic potentiation at the neuromuscular junction in buccal muscle. The degree of post-tetanic potentiation is a function of the frequency of spikes within bursts of motor neuron activity, and of the interburst interval. The MCC may affect both these parameters and thus can indirectly affect EJP size. In addition, the MCC can act directly at the muscle to affect the rate of build-up and the final value of the force of contractions elicited by successive bursts of spikes in motor neurons. Because of the large number of relevant variables and their interaction, a quantitative assessment of the relative role of the different variables in producing arousal has not been done and may prove difficult to accomplish.

Although we have uncovered a number of physiological variables that are likely to be important in food arousal in *Aplysia,* we have only been looking at a limited aspect of the problem, and the complete set of relevant mechanisms will very likely prove to be even more complex than is indicated by the present data. We have concentrated largely on the biting response and the role of the MCC. As outlined in the beginning of this paper, food arousal involves several different types of feeding behaviors and can also involve reflexes other than those directly related to feeding. Based on its known connections and axonal distribution, it is unlikely that the MCC plays an important role in the arousal of behaviors other than biting. Furthermore, the MCC itself appears to receive a tonic excitatory input during feeding stimuli. Consequently, we feel that feeding may activate a central arousal system that in turn affects a large number of response systems. The effects of this arousal system on biting are executed by the MCC. Other similar cells may perform a similar function for other response systems. The data on *Aplysia* clearly suggest that the multiple expressions of arousal involve multiple neuronal systems.

The apparent complexity of arousal in *Aplysia* at first may appear to be discouraging, especially since arousal is a relatively simple form of behavioral plasticity, and *Aplysia* has a relatively simple nervous system. Nevertheless, there are signs that underlying this complexity there may be certain simplifying general principles. For example, at the molecular level, diverse mechanisms may all involve cAMP as an intermediary. There are good data suggesting that even other forms of response enhancement in *Aplysia,* such as sensitization of the gill-withdrawal response following a noxious stimulus, involve cAMP and a biogenic amine (Brunelli et al., 1976). Recent evidence from other invertebrates suggests that biogenic amines also may be mediating modulatory synaptic actions related to various forms of behavioral arousal (Evans, Kravitz, and Talamo, 1976*a*; Evans, Kravitz, Talamo, and Wallace, 1976*b*; Hoyle and Dagan, 1978). In the vertebrate nervous system, arousal also appears to involve biogenic amines and cAMP, and as in *Aplysia*, vertebrate arousal involves central as well as peripheral actions. It remains to be determined whether the parallels between arousal in *Aplysia* and in higher animals are adventitious or, instead, point to certain unifying principles that apply across species lines.

ACKNOWLEDGMENTS

We thank K. Hilten for the preparation of figures. Supported in part by National Institute of Neurological and Communicative Disorders and

Stroke grant NS 12492, grant 1 PO1 GM 23540, Scope D, and by an Alfred P. Sloan Foundation training grant.

REFERENCES

Advokat, C., T. Carew, and E. Kandel (1976). Modulation of a simple reflex in *Aplysia californica* by arousal with food stimuli, *Abstr. Annu. Meet. Soc. Neurosci.*, 6th, Toronto, p. 313.

Berry, M. S. and V. W. Pentreath (1976). Properties of a symmetric pair of serotonin-containing neurones in the cerebral ganglia of *Planorbis*, *J. Exp. Biol.* **65**:361–380.

Brunelli, M., V. Castellucci, and E. R. Kandel (1976). Synaptic facilitation and behavioral sensitization in *Aplysia*: possible role of serotonin and cyclic AMP, *Science* **194**:1178–1180.

Cedar, H., E. R. Kandel, and J. H. Schwartz (1972). Cyclic adenosine monophosphate in the nervous system of *Aplysia californica*. I. Increased synthesis in response to synaptic stimulation, *J. Gen. Physiol.* **60**:558–569.

Cedar, H. and J. H. Schwartz (1972). Cyclic adenosine monophosphate in the nervous system of *Aplysia californica*. II. Effect of serotonin and dopamine, *J. Gen. Physiol.* **60**:570–587.

Cohen, J., K. R. Weiss, and I. Kupfermann (1974). Physiology of the neuromuscular system of buccal muscle of *Aplysia*, *Physiologist* **17**:198.

Cohen, J. L., K. R. Weiss, and I. Kupfermann (1978). Motor control of buccal muscles in *Aplysia*, *J. Neurophysiol.* **41**:157–180.

Cottrell, G. A. (1970). Direct postsynaptic responses to stimulation of serotonin-containing neurones, *Nature (London)* **225**:1060–1062.

Cottrell, G. A. and J. B. Macon (1974). Synaptic connexions of two symmetrically placed giant serotonin-containing neurones, *J. Physiol. (London)* **236**:435–464.

Dieringer, N., J. Koester, and K. R. Weiss (1978). Adaptive changes in heart rate of *Aplysia californica*, *J. Comp. Physiol.* **123**:11–21.

Eisenstadt, M., J. E. Goldman, E. R. Kandel, H. Koike, J. Koester, and J. H. Schwartz (1973). Intrasomatic injection of radioactive precursors for studying transmitter synthesis in identified neurons of *Aplysia californica*, *Proc. Natl. Acad. Sci. U.S.A.* **70**:3371–3375.

Evans, P. D., E. A. Kravitz, and B. R. Talamo (1976a). Octopamine release at two points along lobster nerve trunks, *J. Physiol. (London)* **262**:71–89.

Evans, P. D., E. A. Kravitz, B. R. Talamo, and B. G. Wallace (1976b). The association of octopamine with specific neurones along lobster nerve trunks, *J. Physiol. (London)* **262**:51–70.

Frings, H. and C. Frings (1965). Chemosensory bases of food-finding and feeding in *Aplysia juliana* (Mollusca, Opisthobranchia), *Biol. Bull. Woods Hole* **128**:211–217.

Gelperin, A. and J. J. Chang (1976). Molluscan feeding motor program: response to lip chemostimulation and modulation by identified serotonergic interneurons, *Abstr. Annu. Meet. Soc. Neurosci.*, 6th, Toronto, p. 322.

Gillette, R. and W. J. Davis (1977). The role of the metacerebral giant neuron in the feeding behavior of *Pleurobranchaea*, *J. Comp. Physiol.* **116**:129–159.

Gillman, A. G. (1970). A protein binding assay for adenosine 3':5'-cyclic monophosphate, *Proc. Natl. Acad. Sci. U.S.A.* **67**:305–312.

Granzow, B. and S. B. Kater (1978). Identified higher-order neurons controlling the feeding motor program of *Helisoma, Neuroscience* **2**:1049–1063.

Hoyle, G. and D. Dagan (1978). Physiological characteristics and reflex activation of DUM (octopaminergic) neurons of locust metathoracic ganglion, *J. Neurobiol.* **1**:59–79.

Kandel, E. R., M. Brunelli, J. Byrne, and V. Castellucci (1976). A common presynaptic locus for the synaptic changes underlying short-term habituation and sensitization of the gill-withdrawal reflex in *Aplysia, Cold Spring Harbor Symp. Quant. Biol.* **40**:465–482.

Kandel, E. R. and L. Tauc (1966). Anomalous rectification in the metacerebral giant cells and its consequences for synaptic transmission, *J. Physiol. (London)* **183**:287–304.

Kater, S. B., C. Heyer, and J. P. Hegmann (1971). Neuromuscular transmission in the gastropod mollusc *Helisoma trivolvis*: identification of motoneurons, *Z. Vgl. Physiol.* **74**:127–139.

Kupfermann, I. (1974). Feeding behavior in *Aplysia*: a simple system for the study of motivation, *Behav. Biol.* **10**:1–26.

Nabias, B. de (1894). Récherches histologiques et organologiques sur les centres nerveux de gastéropodes, *Actes Societé Linnae (Bordeaux)* **47**:1–202.

Orkand, P. M. and R. K. Orkand (1975). Neuromuscular junctions in the buccal mass of *Aplysia*: fine structure and electrophysiology of excitatory transmission, *J. Neurobiol.* **6**:531–548.

Paupardin-Tritsch, D. and H. M. Gerschenfeld (1973). Transmitter role of serotonin in identified synapses in *Aplysia* nervous system, *Brain Res.* **58**:529–534.

Preston, R. J. and R. M. Lee (1973). Feeding behavior in *Aplysia californica*: role of chemical and tactile stimuli, *J. Comp. Physiol. Psychol.* **82**:368–381.

Senseman, D. and A. Gelperin (1974). Comparative aspects of the morphology and physiology of a single identifiable neuron in *Helix aspersa, Limax maximus* and *Ariolimax californica, Malacol. Rev.* **7**:51–52.

Susswein, A. J., K. R. Weiss, and I. Kupfermann (1978). The effects of food arousal on the latency of biting in *Aplysia, J. Comp. Physiol.* **123**:31–41.

Weinreich, D., M. W. McCaman, R. E. McCaman, and J. E. Vaughn (1973). Chemical, enzymatic and ultrastructural characterization of 5-hydroxytryptamine-containing neurons from ganglia of *Aplysia californica* and *Tritonia diomedia, J. Neurochem.* **20**:969–976.

Weiss, K. R., J. Cohen, and I. Kupfermann (1975). Potentiation of muscle contraction: a possible modulatory function of an identified serotonergic cell in *Aplysia, Brain Res.* **99**:381–386.

Weiss, K. R., J. L. Cohen, and I. Kupfermann (1978). Modulatory control of buccal musculature by a serotonergic neuron (metacerebral cell) in *Aplysia, J. Neurophysiol.* **41**:181–203.

Weiss, K. R. and I. Kupfermann (1976). Homology of the giant serotonergic neurons (metacerebral cells) in *Aplysia* and pulmonate molluscs, *Brain Res.* **117**:33–49.

Weiss, K. R., M. Schonberg, J. Cohen, D. Mandelbaum, and I. Kupfermann (1976). Modulation of muscle contraction by a serotonergic neuron: possible role of cyclic AMP, *Abstr. Annu. Meet. Soc. Neurosci.*, 6th, Toronto, p. 338.

CONTROL OF SLEEP-WAKING STATE ALTERATION IN *FELIX DOMESTICUS*

Robert W. McCarley

Harvard Medical School, Boston, Massachusetts

INTRODUCTION

The behavioral state of desynchronized sleep is present in all placental and, with one exception, in all marsupial mammals that have been studied (Van Twyver and Allison, 1970). In humans it is associated with the psychological phenomenon of dreaming, and in all species desynchronized sleep (D sleep) has the attractive features of periodic occurrence and rather constant physiological and behavioral manifestations. In this chapter, I will note its several advantages as a phenomenon for analysis by the student of behavior, will survey cellular evidence bearing on the generation of the events of this state, and will present a hypothesis about the mechanism of its periodic occurrence.

Criteria for the selection of a behavior as the object of physiological analysis might include the following:

—The behavior should be readily, reliably, and objectively defined.
—The behavior should be phylogenetically conservative so that a number of animal species could be used as subjects.
—The behavior and the associated physiological events should be consistent from occurrence to occurrence.
—The behavior should be of sufficiently long duration and occur sufficiently often that its physiological basis can be explored in experiments.
—The behavior should not be disruptive of the physiological experimental techniques used to explore it.

For maximal relevance one might add:

—The behavior should be of intrinsic interest and importance for humans and/or be present in humans.

Desynchronized sleep has all these advantages, as will be seen in the following discussion. In particular, the constancy of its features from occurrence to occurrence deserves emphasis. I must note three disadvantages to its study. First, it is not under experimental control and the experimenter cannot elicit it at will but instead must wait on the animal's clock. The second is that while there have been great strides in understanding mechanisms in recent years, the function(s) of desynchronized sleep remains obscure. Nonetheless, its ubiquity over species and the percentage of time it occupies suggest, from an evolutionary point of view, that it will be found to subserve important biological functions. The third disadvantage is that it is a state occurring in a complex brain, making more difficult the application of cell-to-cell analysis of connectivity and influence. This point is at least partially offset by the relative simplicity of the state and of its behavior and physiology.

From the standpoint of behavior, desynchronized sleep in the cat is characterized by a recumbent, curled posture with flaccid musculature upon which are superimposed muscle and vibrissae twitches. As observed in antiquity by Lucretius (44 B.C.) and others, animals, especially dogs, may have movements of the limbs and associated vocalizations during this phase of sleep "as if they were hunting." However, such behavioral manifestations are neither definite enough nor reliable enough to offer a firm basis for defining the state. Thus, most definitions of the state proceed from a different set of criteria, those derived from the first investigations of this state which used macropotential or electroencephalographic (EEG) techniques and the ink-writing oscillograph for records. These are accordingly called the *electrographic* criteria and are of importance because they reproducibly and reliably define this particular behavioral state. These electrographic findings and the results from recordings of neurons during desynchronized sleep comprise the phenomena that any theory of the occurrence of desynchronized sleep must explain. They are:

Muscular: Muscle atonia, especially nuchal and antigravity muscles. Phasic or short-duration twitches of muscles, especially of distal flexors.

Extraoculomotor/Visual System: The occurrence of rapid eye movements. The occurrence of EEG spikes in the pons, lateral geniculate nucleus, and visual cortex (ponto-geniculo-occipital or PGO waves).
Cortical: Desynchronized (actually low-voltage fast) electroencephalographic pattern.
Temporal: The occurrence of the desynchronized phase of sleep on a periodic basis.
Cellular Activity: Generalized high discharge rates of cells.

In this chapter I shall sketch a model for the cellular generation of these phenomena of desynchronized sleep. As a preface to this discussion, it is important to have an explicit list of criteria for cellular generators of events, whether these events are cellular, electrographic, behavioral acts or behavioral states. I shall use the word "event" as a shorthand for all these classes of observations, whether the increased cellular discharge of D sleep, the rapid eye movements of D sleep, the muscle twitches, or the behavioral state of D sleep itself. The following list of generator criteria should be used *in ensemble* for determining the most likely cellular candidates for a generating or controlling role in physiological and behavioral events.

Criteria for Cellular Generators of D Sleep Phenomena

(1) Selectivity of modulation

(a) If the discharge of the cells is excitatory, i.e., postulated to elicit the event, then these cells should show a concentration of discharge in conjunction with the event or state they control. (b) If the cells are suppressive, i.e., if their discharge suppresses the event, then the neurons should show decreased discharge in association with the state. (c) The expected discharge characteristics of generator cells at times other than during the putative controlled event involve a consideration of whether these cells participate in other events, either as generator or follower cells. If, for example, the cells involved in generation of the event do not participate in other events, then their discharge rate during times other than during the controlled event may be expected to be low, and selectivity of modulation may simply be measured as the ratio of discharge rate during the controlled event to that at other times, i.e., selectivity of modulation can be measured as the selectivity of discharge and determined in extracellular recordings. It is of obvious advantage to begin the study of the generators of an event with experimental conditions allowing as much isolation as possible.

In the most general sense, the concept of selectivity of modulation refers to event-specific changes in membrane properties, such as alterations in membrane impedance and potential, that affect the propensity of a cell to discharge. It is to be expected that generator cells for long-duration events such as behavioral states will show a long-duration or tonic modulation of their membrane properties in conjunction with the controlled event or state. For example, generator cells for desynchronized sleep will be those that show the maximum D sleep-specific tonic modulation of membrane properties, as measured by ratios of values during D sleep to those at other times.

Measurement of the modulation of membrane properties in addition to the simple measurement of discharge frequency modulation is most useful and important when neurons show discharges in association with more than one event. For example, consider the case of cells postulated to act as generators for a long-duration behavioral state, but which also show short-term or phasic discharges in association with one or more short-term or phasic events outside the putative controlled state and thus may also be acting as either generators or follower cells for these short-term events. Determination of the presence or absence of state-specific, tonic modulation of membrane properties can be made rather easily by intracellular measurements, but if cellular discharge is also associated with other events, then there will be no state-specific discharge, and selectivity of discharge alone will not be as helpful as a generator criterion. This point will be discussed further, using desynchronized sleep as an example.

(2) Temporal relationships

Generator cells should show alterations in discharge that precede the events they are said to control. (a) The phasic latency is a measure of the duration by which discharge alterations precede the short-duration or phasic events. Clearly the earliest lead time marks the best candidate for the generator of the event. (b) The tonic latency is a measure of the phase lead of cell groups for the tonic or long-duration events; for the case of desynchronized sleep, it is the lead time for the state itself. Again, the longest tonic latency is the best candidate, in terms of this criterion, for being the generator of the state.

(3) Repeated association

The discharge of generator cells should show repeated association with the events they are said to control; thus in the case of periodic events, the generator cells should show periodicity.

(4) Connectivity

Generator cells should have (a) the requisite connectivity to effector cells and (b) the correct sign of influence (excitatory or suppressive).

CONTROL OF THE DESYNCHRONIZED PHASE OF SLEEP

I now turn from general questions about cellular criteria for control of events to a discussion of specific evidence relevant to the control of the desynchronized phase of sleep.

The first group of investigations seeking areas of the brain critical for generation of the desynchronized phase of sleep employed lesion and stimulation techniques; these investigations strongly implicated the pontine brainstem as critical for desynchronized sleep phenomena. (See Moruzzi, 1972, for a comprehensive review focused on these techniques.) Perhaps the most interesting of the lesion preparations was the "pontine cat," a chronic cat with a prepontine transection that preserved the pons and bulbar brainstem caudal to the cut, but destroyed brain rostral to the pons. The pontine cat had periodic episodes of muscle atonia, spikey waves in the pons (the pontine PGO waves), and some rapid eye movements (VI is preserved), all with a periodicity and duration like that of D sleep episodes in the normal cat. The clear implication was that the basic generative mechanisms and source of periodicity are contained in the pons or bulb. Since it was pontine tegmental lesions that strongly suppressed D sleep, the further conclusion was that the pontine structures were necessary and sufficient for the occurrence of D sleep.

Discharge Rate Changes During Desynchronized Sleep

Thus cellular recordings in the pons are of special interest. The first recordings in sleep were made in unanesthetized but head-restrained cats, and showed that cells in the pontine reticular formation and particularly in the gigantocellular tegmental field (FTG; Berman, 1968) underwent dramatic modulations of discharge activity over the sleep-waking cycle (Figures 1, 2, and 3; McCarley and Hobson, 1971; Hobson, McCarley, Pivik, and Freedman, 1974*b*). (It should be appreciated that the FTG includes both the nucleus reticularis pontis oralis and caudalis in Olszewski and Baxter's [1954] terminology.)

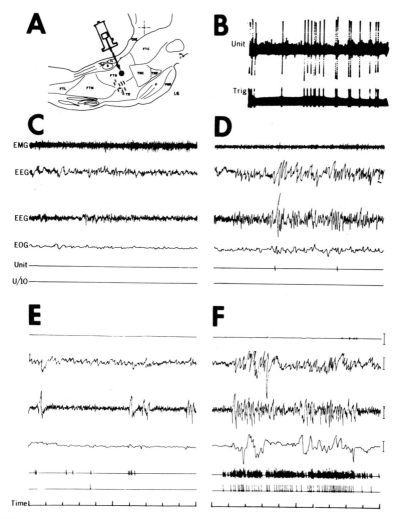

FIGURE 1. Selectivity of discharge. In A, the schematic of the microelectrode attached to the micromanipulator indicates the histologically localized recording site in the giant cell field (FTG; see Figure 2 for other abbreviations). In B, 10 sec of extracellular microelectrode recording are shown together with pulses (triggered by the action potentials of the unit) that were used in computer data analysis. Parts C through F show the polygraph records of unit activity in each of four behavioral states. In C, waking, the unit is silent. In D, synchronized sleep, sporadic firing occurs. In E, desynchronized sleep without rapid eye movements (REM), firings are more abundant and tend to occur together with PGO wave activity in the EEG. In F, also desynchronized sleep, firings are most intense coincident with a burst of REMs and intense PGO wave activity. Note that unit activity begins to increase before the first eye movement and subsides gradually after the last eye movement. Modified from Hobson et al. (1974b).

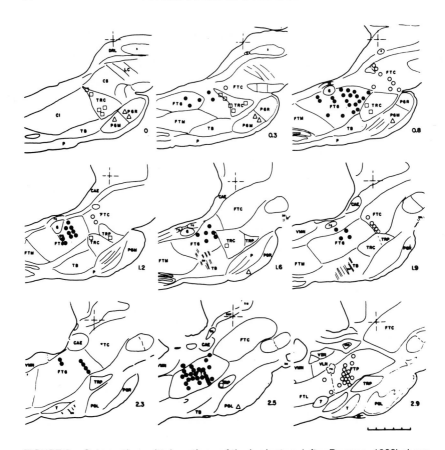

FIGURE 2.   Schematic sagittal sections of the brainstem (after Berman, 1968) show the location of 130 recording sites used for selectivity, discharge pattern, time course, and eye movement data. Sections are taken progressively more laterally from the midline; 0 mm is upper left and 2.9 mm is lower right. Calibration scale is in millimeters. The center lines in the upper right of each schema indicate the intersection of coordinates AP_0 and HC_0. Cells encountered at recording sites shown were assigned to one of four nuclear groups as follows (see anatomical key below): 74 FTG neurons, solid circles; 32 FTC, FTP, and FTL neurons, open circles; 16 TRC and TRP neurons, open squares; 8 pontine gray (PGM, PGR, and PGL) neurons, open triangles. The large number of recording sites at 0.8 and at 2.5 is a function of the arbitrarily planned position of penetrations. Note that most of the FTG loci are in the rostral part of that field (anterior to the abducens nucleus).

Anatomical key: CAE, locus coeruleus; CI, centralis inferior (raphe); CS, centralis superior (raphe); DRL, dorsalis rostralis lateralis (raphe); FTC, central tegmental field; FTG, gigantocellular tegmental field; FTL, lateral tegmental field; FTM, medial

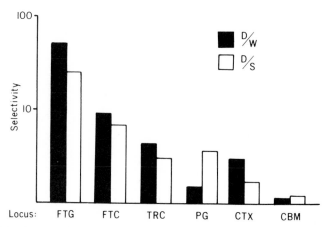

FIGURE 3. Selectivity of discharge as a function of anatomical location. The ratios of geometric mean rates in D to those in W (solid bars) and to those in S (open bars) are shown (on a log scale) for four brainstem nuclei and two cortical cell populations. Selectivity is maximal in the giant cell zone (FTG) and declines progressively in the adjacent tegmental fields (FTC, FTL, and FTM). The nonreticular nucleus of Bechterew (TRC) and pontine gray (PG) have values comparable to those of cerebral cortical neurons in the lateral gyrus (CTX). Cerebellar Purkinje cells (CBM) are very unselective.

Neurons in this area had discharge rates that increased some 40- to 50-fold as the animal passed from synchronized to desynchronized sleep; Figure 3 shows that this discharge selectivity (the discharge rate in D sleep compared with synchronized sleep [S] and waking [W]) was maximal in the giant cell field when compared with other pontine reticular regions and with other brain recording sites. This selectivity of discharge suggested that these cells should be examined more closely for a possible generating role in D sleep.

tegmental field; FTP, paralemniscal tegmental field; LC, linearis centralis (raphe); P, pyramidal tract; PGL, pontine gray, lateral division; PGM, pontine gray, medial division; PGR, pontine gray, rostral division; T, nucleus of trapezoid body; TB, trapezoid body; TRC, tegmental reticular nucleus, central division; TRP, tegmental reticular nucleus, pericentral division; VLN, lateral vestibular nucleus; VMN, medial vestibular nucleus; VSN, superior vestibular nucleus; 3, oculomotor nucleus; 3N, oculomotor nerve; 4, trochlear nucleus; 5M, mesencephalic trigeminal nucleus; 5MT, mesencephalic trigeminal tract; 6, abducens nucleus; 6N, abducens nerve; 7, facial nucleus; 7G, genu of facial nerve; 7N, facial nerve. Modified from Hobson et al. (1974b).

Discharge Rate Changes During the Phasic Events of Desynchronized Sleep

A detailed autocorrelational analysis (McCarley and Hobson, 1975*a*) of the discharge pattern of the giant cell field cells showed that these cells tended to discharge in runs of clusters of activity (Figure 4); a natural question was whether these clusters of discharge activity were correlated with the phasic or short-duration events of D sleep such as REM and PGO waves. Accordingly, the relationship between the onset of isolated eye movements in D sleep and the discharge of FTG and other pontine cell groups was quantified (Figure 5). Cross-correlation techniques (Pivik, McCarley, and Hobson, 1977) showed that FTG cells have changes in their discharge rates that begin at least 100–150 msec before eye movement onset and have, of the pontine brainstem populations examined, the most consistent, earliest, and most prominent discharge rate increases. Inflection points for discharge rate increases in FTG cells peaked at 150–100 msec before the onset of eye movements, and discharge maxima occurred in the 50 msec before and after eye movement onset. Thus, there is considerable correlational evidence that cells in the pontine giant cell field may act as generators for the rapid eye movements of desynchronized sleep.

It is important to note that many of these FTG cells were recorded in

---

FIGURE 4. Desynchronized sleep discharge patterns. Oscilloscope photographs (A) and raster displays (B) of a phasically discharging FTG unit and a tonically discharging FTC unit. Oscilloscope sweep duration is 1 sec. The raster display represents each neuronal discharge as a point and is to be read left to right and top to bottom; each row is 1.3 sec duration, and about 90 sec of data are represented on the whole display. Neither the oscilloscope photographs nor the raster displays show any strong regularity or stereotypy of interspike interval durations for either unit. Note, however, the marked differences between units in the degree of sustained versus clustered discharges on the raster display. C and D are autocorrelograms derived from stationary portions of the spike trains of the tonically and phasically discharging units. Part C displays the long-lag duration (6.4 sec maximum; 64 msec bin width) autocorrelograms. Ordinates are number of counts/100 in each bin. Note the flatness of the autocorrelogram from the tonic unit and the prominent autocorrelogram peaking in the phasic unit. Part D displays autocorrelograms with a much smaller maximum lag (80 msec) and bin width (0.8 msec). Ordinate is number of counts in each bin. No rhythmicity is evident. Rhythmicity of discharge or regularity of interspike intervals, were it present, would appear as a regular succession of peaks and valleys of high and low probability of discharge. This is not seen. FTG cells tend to discharge in irregular runs or clusters of spikes. After McCarley and Hobson (1975*a*).

spinal cord. The chief spinal cord projection pathway for pontine neurons is the medial reticulospinal tract (RSTm); this tract is predominantly excitatory to spinal cord motoneurons. Microstimulation studies have shown that reticulospinal axons have extensive branching within the cord, often projecting to more than one segment and showing extensive branching within a segment (Peterson, Maunz, Pitts, and Mackel, 1975). Thus, neurons projecting to the spinal cord from the pontine reticular formation have the requisite density, extent, and sign of influence to mediate the phasic muscle twitches occurring in D sleep. A recent recording study of antidromically identified reticulospinal neurons during D sleep (Wyzinski, McCarley, and Hobson, 1978) shows that antidromically identified reticulospinal cells have the requisite discharge characteristics, since they discharge in association with the phasic events of D sleep, including the muscle twitches.

An interesting point about function is raised by this study, since many of the cells identified as reticulospinal were recorded in the PPRF and had discharges correlated with D sleep eye movements. The studies of PPRF cell discharge in relation to waking eye movements that were discussed earlier did not test whether these eye movement-related cells also had spinal projections. However, both the extensive anatomical and physiological studies on reticulospinal projections cited earlier and the Wyzinski, Hobson, and McCarley (1978) data suggest the likelihood of a spinal projection of the PPRF cells whose discharge is correlated with waking eye movements. The experimental finding of such a projection would indicate that these cells might act to coordinate oculo- and somatomotor events, as has been proposed for more caudal pontoreticular cells (Peterson, 1977).

Tonic Latency

I have thus far considered evidence about the association of pontine reticular cell discharge with some of the phasic events of D sleep. It is significant, in terms of the tonic latency generator criterion applied to the onset of D sleep periods themselves, that pontine reticular cells showed discharge rate increases that preceded any EEG signs of D sleep (Hobson, McCarley, Freedman, and Pivik, 1974*a*). Extremely conservative statistical analysis indicated that many pontine reticular cells significantly increased discharge rate 3–5 min before D sleep onset (Figure 7). This longest anticipatory discharge rate increase might be expected in cells whose activity was responsible for generation of the

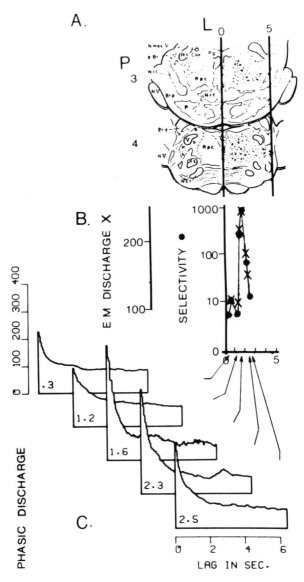

FIGURE 6. Desynchronized sleep discharge selectivity, eye movement-related discharge, and phasic or clustered discharge are maximal in the center of the giant cell field. In A, two frontal sections of the pontine brainstem as drawn by Brodal (1957) and corresponding to the sagittal sections in Figure 2 are reproduced to show the distribution of cell sizes within the FTG (R.p.c. in Brodal's drawing). We have superimposed a midline and a line 5 mm lateral to it to correspond to the abscissa in B. In B, the ordinates are: FTG population selectivity of discharge in D sleep (circles)

events of the D sleep episodes; presumed "follower" cells in occipital cortex did not have this long anticipatory discharge rate increase. For forming a hypothesis about the process(es) involved in the marked increase in pontine reticular firing rate with the onset of D sleep, the discovery that the form of the recruitment curve is exponential (Figure 7C) is helpful. This indicates that the number of cells attaining a higher discharge level is proportional to the number of cells that have already done so. This is compatible with a model postulating that, at this point in the sleep cycle, there is a positive feedback operating on members of the reticular population. The strong histological evidence for recurrent collaterals (see Table 1) and the report of Ito, Udo, and Mano (1970) of physiological evidence that such collaterals are excitatory lend support to this hypothesis.

Periodicity

The long anticipatory discharge rate increase before D sleep onset of FTG cells is clearly evident even in raw data plots of discharge time course obtained in recordings lasting many hours and covering many sleep-waking cycles. The plot in Figure 8 shows that these units have a consistent, periodic trajectory of discharge over the sleep-waking cycle, and this repeated correlation between unit discharge and the sleep cycle period satisfies the periodicity criterion for generator cells. Their discharge shows peaks in D sleep, a rapid decline just before D offset to a nadir usually associated with waking, a slow increase in synchronized sleep, and an explosive acceleration at D sleep onset.

To summarize the evidence thus far presented: The discharge rate increases of pontine reticular cells that occur long before the onset of the eye movements and PGO waves of D sleep are correlative evidence that these cells may serve as generators for these events. Similarly there is a

---

and, also for the FTG population, the relative heights of the D sleep peaks of eye movement-associated discharge (asterisks) as compared with average discharge levels for a time period 500 msec before and after eye movement onset. Note that discharge selectivity and eye movement-related discharge are highly correlated with each other and that both appear maximal in giant cell-rich portions of the field. Panel C shows that the degree of phasic or clustered discharge roughly parallels selectivity and eye movement-related discharge and is also maximal in giant cell-rich portions of the field. Ordinate is the percentage of counts expected in each bin of an autocorrelogram with no peaking and obtained by bin-by-bin averaging of FTG autocorrelograms; the height and duration of the autocorrelogram peak are proportional to the intensity and duration of runs of clustered discharge. (Other autocorrelogram parameters are as in Figure 3.) Modified from Hobson et al. (1974*b*); McCarley and Hobson (1975*a*); Pivik et al. (1977).

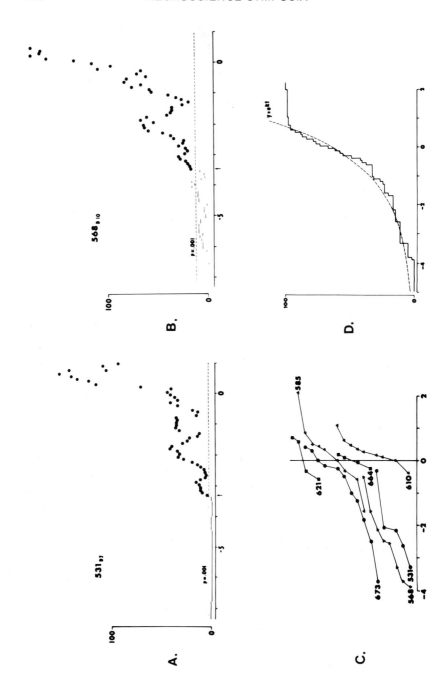

discharge rate increase anticipating the onset of desynchronized sleep itself, also a phenomenon to be expected in generator cells, as well as consistent periodic time course. Obviously such correlative evidence is not proof, but it does offer rather strong *prima facie* evidence that these giant cell field cells may be important in the events of D sleep. The important question of what is responsible for the periodic modulation of FTG cell discharge and for the periodic occurrence of D sleep itself will be considered later.

Inverse Selectivity of Discharge: Neurons Recorded in the Locus Coeruleus and Raphe

There are two populations of brainstem cells with the important property of inverse discharge selectivity in D sleep, i.e., markedly lower

---

FIGURE 7.   Tonic latency. Panels A and B: Determination of tonic latency for a transition period. Two examples, the tenth desynchronized sleep period from unit 568 and the second desynchronized sleep period from unit 531, are shown. Discharge level, defined as the number of spikes in the preceding 30 sec and shown as ordinates, was calculated at 5-sec intervals, starting 7 min before D sleep onset (time 0 shown as abscissas). The discharge level, which is significantly different from the baseline period (P = 0.001), is shown as a dashed line. Thus, the tonic latency of transition for unit 568 was 210 sec before D sleep onset; for unit 531, it was 200 sec.

Panel C: Individual cumulative histograms of tonic latency for five FTG and two FTC neurons (units 531 and 673). Tonic latencies were determined for each of 60 transition periods in seven neurons recorded for 4–12 cycles. The set of tonic latencies of each neuron is displayed as a cumulative histogram on a time base relative to D sleep onset. The ordinate is the number of transition periods (calibration = 1 transition). Levels of the starting point for the curve of each cell are arbitrary. The abscissa is time in minutes. Each curve, therefore, is proportional to the likelihood of a cell having shown a significant discharge rate change as a function of time relative to D sleep onset. When more than one transition period for a given cell had the same tonic latency, the ordinate was incremented proportionally, but only one point is shown. For example, the curve of unit 610 presents 12 values as 9 points because there were 3 pairs of identical tonic latencies.

Panel D: Pooled cumulative histogram of tonic latency for the five FTG and two FTC neurons shown in Panel B. Data from these neurons recorded for multiple cycles were combined to gain insight about the behavior of the population during the transition period. Sixty transition periods from the seven cells are included. The tonic latencies from each transition period were plotted as a cumulative histogram, with a time base relative to D sleep onset. Each latency was weighted so that the contribution from each cell was equal, and the total was set at 100%. The experimental curve (solid line) is shown together with a plot of $y = e^{kt}$, the exponential growth curve (dashed line). The correspondence is quite close until 1 min after D sleep onset, when the experimental data reach an asymptote, indicating that all cells have changed discharge rate. Modified from Hobson et al. (1974a).

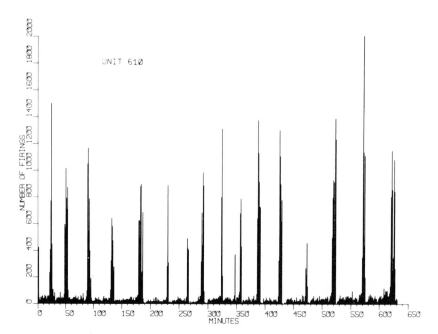

FIGURE 8. Discharge activity of FTG neuron 610 recorded over multiple sleep-waking cycles. Each peak corresponds to a D sleep episode, and a regular trend of discharge activity over a cycle is observable: a peak in D sleep; a rapid decline at the end of the D sleep episode; a trough, often associated with waking; a slow rise (in synchronized sleep); and an explosive acceleration at D sleep onset. Note also the extreme modulation of activity and the periodicity. From McCarley and Hobson (1975*b*).

movements, especially head on neck movements. Different cells appear to discharge preferentially during different kinds of movements. These studies have confirmed the results of high discharge rate ratios for D sleep as compared with synchronized sleep and quiet waking, but the selectivity of discharge as measured by the ratio of maximal D sleep discharge rate to maximal waking discharge rate was only 1.5, and as measured by average D sleep discharge divided by maximal waking discharge was 0.41.

These studies appear to have recorded cells at more caudal pontine levels than those I have discussed (compare Figure 4 in Siegel et al., [1977] and Figure 3 in Vertes [1977] with Figure 2 of this paper) and have not provided the specific information necessary to link waking discharge either to motor events or to their sensory consequences. Nonetheless, these studies are important in that they indicate that FTG cell discharge cannot be regarded as state-specific in the freely moving animal. The absence of state-specificity does not mean that FTG neurons cannot be generators of D sleep events, but raises the question, to be discussed in the next section, of whether FTG neurons are activated in the same manner in both states. It should be emphasized that, to date, the discharge rate decreases of cells in the LC and raphe have not been noted to occur other than in the state of D sleep, even in recordings done in freely moving animals. Thus this discharge rate modulation still appears to be state-specific, although it would not be surprising to find it occurring under some waking conditions.

Theories of Desynchronized Sleep Generation and Neuronal Modulation over the Sleep Cycle

There are two general kinds of hypotheses about the significance of FTG cell discharge during waking and D sleep and its modulation over the sleep cycle. The first class of hypothesis proposes that FTG cells are acting as a type of premotor cell that is part of an executive network for some of the events of D sleep and for certain movements, but that these cells do not play a critical role in the initiation of either of these. FTG cells are conceived to be effector cells, "downstream" to the series of motor commands initiated by some other group(s) of cells. I shall call this hypothesis the simple follower cell model, since it suggests that this is the role of FTG cells during both sleeping and waking. This model proposes that FTG cells discharge during waking because they are "following" discrete excitatory inputs from upstream cells; in D sleep the same process is occurring in FTG cells, but the actual movement is not observed because of the strong inhibition of alpha motoneurons. Unfortunately, it must be observed that the evidence linking FTG cell discharge to specific motor events in waking (with the exception of eye movements), as opposed to sensory consequences of motor events, is largely absent at present and needs to be supplemented with specific correlations between waking motor and muscular events and FTG cell discharge. Such specific correlations have been clearly demonstrated for the eye movements of D sleep, as discussed earlier; the question is whether the FTG cells in D sleep are merely following individual excitatory inputs (the follower cell model) or whether there are other processes (such as disinhibition) at work.

The second class of hypothesis about FTG cell discharge says that a critical component of the increased FTG cell discharge during desynchronized sleep results from tonic, state-specific changes in the excitability of the FTG cells. Both internal (autochthonous) and external (interactive with other cell groups) influences on the excitability of FTG cells should be considered.

The reticular pacemaker model proposes that FTG excitability and discharge are increased during D sleep because of internal metabolic changes that lead to spontaneous membrane depolarization and discharge during this state. Note that the term "pacemaker" is quite specifically used; it refers to the presence of a spontaneous depolarizing drift of the cell membrane during D sleep, just as occurs in pacemaker cells in the heart or in the lobster stomatogastric ganglion (Selverston and Mulloney, 1974; Strumwasser, 1967; Watanabe, Obara, and Akiyama, 1967). This model suggests that FTG cells are generator cells for the events of D sleep, providing the initial impetus, but that during waking they function as follower cells, being driven by other inputs.

The reciprocal interaction model proposes that the increased FTG activity before and during the occurrence of desynchronized sleep episodes is the result of a progressive tonic disinhibition by inhibitory cells that decrease discharge rate with the approach of desynchronized sleep. This model thus proposes that the FTG cells are generator cells for some D sleep events because it is on FTG cells that the disinhibition is acting, and thus it is the FTG cells that undergo a selective modulation of membrane potential. It further postulates that interaction between FTG cells and the inhibitory population forms the basis for the periodic occurrence of D sleep. This model, like the reticular pacemaker model, proposes that FTG cells, although inhibited in waking, can still be driven by inputs and act as follower cells during the activation associated with movements.

There are thus three general classes of hypotheses to explain the observed modulation of FTG cell discharge over the sleep-waking cycle: (1) Endogenous alteration of FTG excitability—the reticular pacemaker model; (2) following of excitatory input—the simple follower cell model; and (3) tonic alteration of excitability because of interaction with an inhibitory cell group—the reciprocal interaction model.

Comments on Selectivity of Modulation as a Generator Criterion

In an earlier statement on criteria for generation of the events of desynchronized sleep, it was proposed that one such criterion was selectivity of discharge during the controlled state, during D sleep (Hobson, 1974). Discharge selectivity was measured by the ratio of discharge rate during D sleep to that during waking and synchronized sleep. This criterion of discharge selectivity thus proposed that discharge must be maximally state-specific or event-specific for a cell to be regarded as a controller of that event.

This criterion was based on the concept that a cell that acted as a generator (an initiator) for one event or state would not be a follower or generator for another state or event. Cellular discharge would therefore be event- or state-specific. This concept now appears to be too narrow. It excludes the possibility that a cell may participate in numerous events, sometimes as a controller, sometimes as a follower, or that it could be part of the controlling network for more than one event. For example, FTG cells discharge in response to novel stimuli, and their extensive spinal, rostral, and intra-brainstem projections are well suited for a role in mediation of components of the startle response. There is some indirect evidence supporting this possibility (Bowker and Morrison, 1976), and hence one would want to leave open the possibility that FTG cells might be generator cells for the startle response in the sense of serving as the critical elements for integration of the sensory stimuli into a motor response. The respiratory and cardiac rhythms provide examples of "generator cells" acting as follower cells during modulation of these events.

Thus the absence of state-specific discharge of FTG cells does not of itself rule out the possibility of FTG cells playing an initiating or controlling role in the events of D sleep. The criterion of selectivity of modulation of membrane properties during the controlled event could be used as a generator criterion; such modulation should be maximal in cells controlling the event.

The nature of the membrane modulation will depend on the precise control mechanism for the event. If, for example, the reciprocal interaction model were accurate, the maximum modulation of membrane potential would be expected to be found in the FTG cells and also, but with an inverse time course, in the inhibitory population of cells with which the FTG cells were interacting. Note that using the criterion of selectivity of modulation means that intracellular recording must be used as a final test of generator selectivity, and that in the case of the postulate of interactive modulation of membrane potential, the observed membrane changes should match those induced by the interacting population of cells. Again I note that it may be strategically useful to investigate the control of an event in experimental conditions affording relative isolation of the event to be studied, and to obtain from these restricted conditions the best first-order candidates for generator cells. This, in fact, is what we did in the first sleep-waking cycle recordings of FTG neurons where head restraint restricted much of the behavior to the sleep-waking cycle.

Evidence Bearing on Theories of Sleep Cycle Control

If the hypothesis of reticular pacemaker cells were correct, one might expect to find evidence of rhythmic, stereotyped discharge in pontine reticular cells. In fact, there is no evidence for this, and the characteristic discharge pattern of FTG cells is runs of clustered discharge with irregular interspike intervals. As Figure 3 shows, this pattern is clearly evident in oscilloscope photographs, raster displays, and in both short- and long-lag autocorrelograms. While intracellular recordings will be necessary because such extracellular evidence cannot rule out the possibility that synaptic input could mask rhythmicity in any pacemaker cells, it should be emphasized that there is no positive evidence of any kind supporting the reticular pacemaker theory.

Figure 9 summarizes the critical conceptual differences between the simple follower cell and the reciprocal interaction model. The reciprocal interaction hypothesis postulates that the time course of FTG membrane potential and excitability is tonically modulated over the sleep cycle by direct interaction with the inhibitory cell population. In this view the FTG cells during waking are inhibited, although it is possible to drive them. During D sleep the FTG cells are disinhibited and thus more likely to discharge to any kind of excitatory input, *including the excitatory collaterals of other FTG cells,* which are critically important for recruitment of population activity.

The simple follower cell hypothesis says that there is no selective tonic modulation of membrane properties or excitability over the sleep cycle; what happens is that FTG cells are driven in waking during particular kinds of movements and that they are similarly driven during desynchronized sleep, although muscle paralysis prevents the expression of movement.

It is apparent that intracellular recordings will be needed for conclusive and direct evidence ruling out or supporting the "follower" cell hypothesis, but there are several lines of evidence already available mitigating against FTG cells simply being "follower" cells in D sleep. As

A. FTG CELL MODULATION – RECIPROCAL INTERACTION

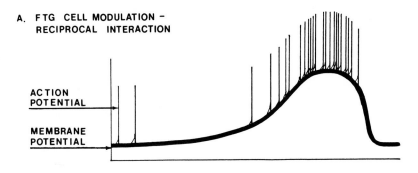

ACTION POTENTIAL

MEMBRANE POTENTIAL

B. NO FTG CELL MODULATION – FOLLOWER CELL

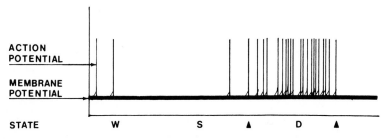

ACTION POTENTIAL

MEMBRANE POTENTIAL

STATE   W   S   ▲   D   ▲

FIGURE 9.   Schematic contrast of time course of FTG cell membrane potential over the sleep cycle as predicted by (A) the reciprocal interaction model and (B) the simple follower cell model. Panel A: The reciprocal interaction model predicts a modulation of the membrane potential because LC and/or raphe cells tonically inhibit FTG cells in waking (W) and synchronized sleep (S) and disinhibit in desynchronized sleep (D) (D onset and offset are indicated by the triangles). This disinhibition and the recurrent collateral excitation from other FTG cells produce the illustrated tonic depolarization in D. Panel B: In the simple follower cell model there is no tonic modulation of membrane potential; rather, there are action potentials arising from individual depolarizing inputs from the driving cells. It can be seen that both models postulate individual depolarizing inputs as the cause of discharge activity during waking. Note that intracellular recording can easily determine if a modulation of membrane potential is present or not. If a modulation is present it can be associated with LC/raphe by showing that the spontaneous changes in membrane properties in W and S are identical to those observed with LC/raphe stimulation.

already stated: (1) The pontine cat preparation and lesion studies indicate that it is the pontine brainstem that is necessary and sufficient for D sleep phenomena. (2) No other cell group recorded in the brainstem (or elsewhere) has the extreme degree of modulation of the

FTG cells and the same temporal course of discharge within D sleep (see summary in Steriade and Hobson, 1976). (3) These pontine reticular cells begin to increase discharge rate during synchronized sleep, several minutes before the occurrence of D sleep episodes, long before any discharge rate changes have been demonstrated in other cell groups. Further, the several minutes before D sleep onset is a time *without muscle atonia but also without movement*. The question here becomes: If the FTG cells are simply "follower" cells, what are they following, since there is neither movement (if they are premotor cells) nor sensory feedback from movement (if they are responding to this input)? I thus turn to an examination of the evidence for the reciprocal interaction model.

THE RECIPROCAL INTERACTION MODEL OF SLEEP CYCLE CONTROL

This model postulates that the basis of the periodic occurrence of D sleep is an interaction between an excitatory cell group (FTG) and an inhibitory cell group(s) that has a time course of discharge activity that is reciprocal to the excitatory cells (McCarley and Hobson, 1975*b*).

Cells with this reciprocal time course were found in microelectrode descents through the locus coeruleus and subcoeruleus (LC); these cells radically decreased discharge rate with the approach of D sleep episodes (D-off cells; Figures 10 and 11). (A minority of LC cells showed a weak discharge rate increase.) During the transition from the synchronized to the desynchronized phase of sleep, the D-off LC cells showed a discharge rate trend opposite to the FTG cells: as LC neuronal activity diminished, FTG activity augmented. Chu and Bloom (1974*b*) have confirmed the finding of D-off LC cells, and their histofluorescent histology suggested that the D-off LC cells are norepinephrine-containing cells (as were D-on cells that increased discharge rate in D sleep, which they found to be more common). Such norepinephrine-containing cells have been shown to have long-lasting inhibitory effects at their synapses in cerebellum and hippocampus (Bloom, Hoffer, and Siggins, 1971; Hoffer, Siggins, and Bloom, 1971; Hoffer, Siggins, Oliver, and Bloom, 1971; Hoffer, Siggins, Oliver, and Bloom, 1973; Oliver and Segal, 1974; Segal and Bloom, 1974*a*; Segal and Bloom, 1974*b*). Further, there is anatomical and physiological evidence of reciprocal connections between the populations of giant cell field cells and the locus coeruleus D-off cells and of recurrent "feedback" connections of each population

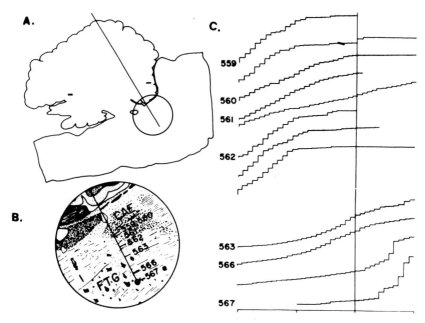

FIGURE 10. Reciprocal discharge by cells in the nucleus locus coeruleus (LC), 559–562, and the gigantocellular field of the anterior pontine tegmentum (FTG), 563–567. A: Outline tracing of a sagittal section of the cat pontine brain at 2.5 mm lateral to the midline showing the path of an exploring microelectrode passing through the cerebellum into the dorsal brainstem. B: Detail of circled area in A, showing the location of the seven successive recording sites in the penetration; circle diameter is 5 mm. C: Cumulative discharge histograms of the cells recorded at the seven sites shown in B during the transition period beginning 2 min before D sleep onset (vertical line) and ending 1 min thereafter. Each activity curve shows the cumulative percentage of discharge for as much of the epoch as was free of arousal. Units recorded in the LC and subcoeruleus slow discharge rate (negatively inflected histograms), while those in the FTG increase discharge rate (positively inflected histograms) with the approach and onset of D sleep.

on itself. Table 1 summarizes these connectivity data. Other evidence indicates that the giant cells are likely to be cholinergic and cholinoceptive and to have generally excitatory effects (see review in Steriade and Hobson, 1976).

Figure 12A presents this structural model of interaction between the FTG population and a population of inhibitory cells. In the figure and in the following discussion of aspects of the model, the LC population is used as the exemplar of an inhibitory population, since it is from these

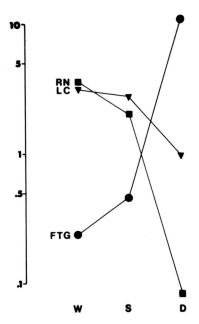

FIGURE 11.   Pooled geometric mean discharge rates for populations of 34 giant cell field (FTG) neurons (circles), 21 locus coeruleus (LC) neurons (triangles), and 18 dorsal raphe nucleus (RN) neurons (squares) during waking (W), synchronized sleep (S), and desynchronized sleep (D). LC and FTG data from Hobson et al. (1975); RN means computed from data in McGinty (1973).

cells that most quantitative data are available. It is to be emphasized that serotonergic dorsal raphe neurons also decrease discharge rate with D sleep onset (McGinty et al., 1973; Figure 11) and may have similar inhibitory roles with respect to FTG cells; these connections are included in Table 1.

It is to be noted that the notion of biogenic amine-containing cells acting to modulate, or set the level for, probability of discharge of other cells, as the reciprocal interaction model proposes for the sleep-waking cycle, is compatible with other authors' formulation of the functional role of these cells (see, for example, Bloom, 1975). It must also be pointed out that the identification of the D-off cells recorded in the raphe with norepinephrine- and serotonin-containing cells, respectively, remains tentative, although the available evidence supports this possibility.

The interactions of the FTG (excitatory) and locus coeruleus

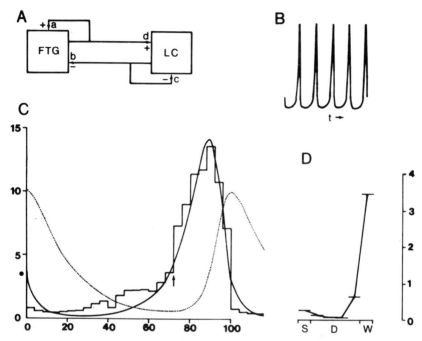

FIGURE 12. A: Structural model of interaction between FTG and LC cell populations. The plus sign implies excitatory and the minus sign inhibitory influences. The letters *a, b, c,* and *d* correspond to the constants associated with the strength of the connections and included in the text equations. B: Theoretical curve derived from the model that best fits the FTG unit in Figure 1. C: The solid-line histogram is the average discharge level of FTG unit 568 over 12 sleep-waking cycles, each normalized to constant duration. The cycle begins with the end of D sleep, and the arrow indicates the bin with the most probable time of D sleep onset. The solid curve describes the FTG fit and the dotted line the LC fit derived from the model. The dot in the ordinate scale indicates the equilibrium values for the two populations. D: Geometric mean values of the discharge activity of 10 LC cells before (synchronized sleep, S), during (D), and after (waking, W) a D sleep episode. Each time epoch is equal to one-quarter of a D sleep period. Note that the discharge rate increase begins in the last quarter of D. From McCarley and Hobson (1975*b*).

(inhibitory) populations can be mathematically modeled by equations of the Lotka-Volterra type, and the time course of activity predicted by the model is in good agreement with actual physiological recordings. According to the model, the events of the sleep cycle can be qualitatively described in this manner: During waking, giant cell activity is low because of tonic inhibition from the LC population of cells. This inhibitory influence gradually decreases because of inhibitory feedback of the LC population on itself. With this disinhibition, giant cell activity increases, at first gradually, then rapidly and exponentially as self-excitation becomes prominent in the giant cell population. This rapid increase of activity marks the onset of a D sleep episode. With the attainment of a high level of giant cell activity, more activity is induced in the LC population; the inhibitory influence of this population ends the high level of giant cell activity and the D phase of sleep, and the cycle starts anew.

The mathematical form of terms describing the influence of each population on itself is suggested by evidence that the rate of change of activity levels in the FTG population is proportional to the current level of activity (Hobson et al., 1974*a*; Pivik et al., 1977). I propose that the same is true for the LC population, but with a negative sign because the recurrent feedback is inhibitory. The highly nonsinusoidal nature of FTG activity suggested that nonlinear FTG-LC interaction was to be expected. This effect is modeled by the simplest form of nonlinearity, the product of activities in the two populations. This is in accord with the reasonable physiological postulate that the effect of an excitatory or inhibitory input to the two populations will be proportional to the current level of discharge activity. Let $x(t)$ be the level of discharge activity in FTG cells, $y(t)$ the level of discharge activity in LC cells, and a, b, c, and d positive constants identified with the strength of the connections outlined in Figure 12A. These terms are related by the equations

$$dx/dt = ax - bxy$$
$$dy/dt = -cy + dxy$$

This system of equations is that of Lotka and Volterra, originally proposed as a model for prey-predator interaction, but also used for chemical systems. In the model, the excitatory (FTG) cells are analogous to the prey population, and the inhibitory (LC) cells are analogous to the predator population. The time course of FTG activity over several sleep-waking cycles predicted by the model is sketched in Figure 9B. (McCarley and Hobson, 1975*b*, describe more mathematical details.)

In Figure 12C a theoretical curve derived from the model is compared with the actual data values for the average of 12 cycles of FTG unit 568. The overall match is rather good. Specifically, both curves show a nadir in the first third of the cycle, a long period of slow growth of activity, and a rapid acceleration as the time of D sleep onset is approached. The average time of D sleep onset occurs at about the same time as the theoretical curve crosses the equilibrium point, and the approach to the peak is less steep than the decline. Comparisons of similarly derived theoretical and observed data curves from other units showed about the same degree of fit (see also Figure 12B). The model predicts that LC activity levels should decline steadily in synchronized sleep to a low point at D sleep onset and then show a rapid rise in the last portion of D sleep (see the theoretical curve in Figure 12C). To determine if the LC pool shows this behavior, the activity curves of 10 D-off LC cells during the successive quartiles of D sleep and for periods of equal duration before and after D sleep were averaged. The sketch of the observed data in Figure 12D is in reasonably good agreement with the theoretical curve in Figure 12C. Note that the increase in discharge activity occurred before the end of the D sleep episode in the averaged data; this important feature was present in each of the 10 LC cells.

There are several comments to be made on this model. One has to do with the way in which one conceptualizes the interrelationships between the LC and raphe populations on the one hand and the FTG population on the other. The raphe and LC populations can be regarded as equivalent in their inhibitory interaction with the FTG population, and their projections to each other can be regarded as cases of the inhibitory feedback the inhibitory population has upon itself (see Table 1 for summary of connectivity evidence). In point of fact, there is a fair amount of evidence that the duration of LC and raphe effects is different: LC neurons act over longer time spans than raphe cells, perhaps by a factor of a hundred or so (see Haigler and Aghajanian, 1974; Bloom, 1975; Segal, 1975). Thus the LC may be the important population for long-duration events, such as the periodicity of D sleep, while short-duration events such as PGO release may be more dependent on raphe effects. With respect to the duration of the sleep cycle, it is apparent that classical IPSP-EPSP durations of the order of milliseconds cannot provide a basis for the twenty-minute-or-so duration of the sleep cycle in the cat. Longer-duration interactive effects are needed; an explicit statement of the nature of such effects is not included in the model, but it can be noted that norepinephrine synaptic effects *are* much longer than the millisecond range of IPSP-EPSPs and extend into the second, perhaps minute, range (Siggins, Oliver, Hoffer, and Bloom, 1971; Hoffer et al., 1973; Segal and Bloom, 1974*a*; Segal and Bloom, 1974*b*).

While the hypothesis that the LC norepinephrine effects are mediated by cyclic AMP (a second messenger) is not without controversy, it does offer a possible mechanism by which a neurotransmitter could alter cellular metabolism and thus have effects whose duration is several orders of magnitude longer than classical EPSP-IPSP effects (Godfraind and Pumain, 1971; Siggins, Hoffer, and Bloom, 1971a, b; Siggins et al., 1971; Bloom, Siggins, and Hoffer, 1974; Lake and Jordan, 1974). The evidence that synaptic activity can trigger long-duration biochemical events has recently been reviewed (Kebabian, Zatz, and O'Dea, 1977; Kanof, Ueda, Uno, and Greengard, 1977); many of these effects occurred at adrenergic synapses and were dependent on cyclic AMP, including a cyclic AMP-mediated effect on transmitter synthesis itself in the postsynaptic cell. While any such effects remain to be investigated and are speculative at the synapses discussed in the reciprocal interaction model, the growing body of evidence about the ubiquity of transmitters with long-duration effects suggest that it is long-duration consequences of transmitter action that most likely form the basis of long-duration behavioral states and rhythms, including desynchronized sleep. In this volume, Weiss and Kupfermann (1978) present evidence suggesting that the long-duration state of food arousal in *Aplysia* depends critically on the cyclic AMP-mediated responses to serotonergic neurons. This kind of model of the basis of long-term effects is to be contrasted with an earlier viewpoint that suggested that long-duration biological rhythms might be the result of beats or resonances resulting from the connections of individual neural oscillators, each operating with a short-duration period (see, for example, Winfree, 1967).

As a further comment on the FTG-raphe interaction, it should also be noted that a recent study by Wang, Gallager, and Aghajanian (1976) of the effect of FTG stimulation on putative serotonergic raphe cells in chloralose-anesthetized rats found evidence for FTG inhibition of these raphe cells, contrary to the model proposed here. Because of the complexities of stimulation effects, especially in anesthetized animals where the only measured effect is extracellularly recorded discharge suppression, these results will need to be confirmed in unanesthetized animals and with intracellular recordings. The latter is necessary (as it is for other synapses postulated in this model) for an exact measure of the PSP latency—i.e., whether it is mono- or polysynaptic—and the PSP nature—i.e., IPSP or disinhibition. Such latency measurements will also be able to rule out raphe recurrent collateral inhibition as a possible cause of the suppressive effects observed.

SUMMARY

I have discussed criteria for cellular generators of behavioral and physiological events and have reviewed evidence suggesting that cells in the gigantocellular tegmental field (FTG) of the pontine reticular formation may act as generators for some of the phasic events of desynchronized sleep, including the eye movements, PGO waves, and muscle twitches. Studies of the relationship of FTG cell discharge to waking eye movements and to body movements suggest that FTG cells may be a critical component of the executive network for these functions also. Pontine reticular cells have discharge rate increases that anticipate the occurrence of desynchronized sleep by several minutes and that show a regular and dramatic cycle of discharge modulation over the sleep-waking cycle. The discharge rate increases during D sleep of this group of cells may result from disinhibition by biogenic amine-containing cells of the locus coeruleus and dorsal raphe, where a majority of cells slow discharge rate with the approach and occurrence of D sleep. The reciprocal interaction model of sleep cycle control postulates that the periodic occurrence of the desynchronized phase of sleep results from the interaction of FTG and biogenic amine-containing cells. Data about the time course of modulation of FTG and locus coeruleus cells are consistent with this model.

ACKNOWLEDGMENTS

I thank Drs. J. Allan Hobson and John P. Nelson for their helpful comments on earlier drafts of this paper. This research was done in collaboration with Dr. Hobson and supported by National Institute of Mental Health grant MH 13,923 and the Milton Fund of Harvard University.

REFERENCES

Aghajanian, G. K. and R. Y. Wang (1977). Habenular and other midbrain raphe afferents demonstrated by a modified retrograde tracing technique, *Brain Res.* **122**:229–242.

Berman, A. L. (1968). *The Brain-Stem of the Cat.* University of Wisconsin Press, Madison.

Bloom, F. E. (1975). The role of cyclic nucleotides in central synaptic function, *Rev. Physiol. Biochem. Pharmacol.* **74**:1–103.

Bloom, F. E., B. J. Hoffer, and G. R. Siggins (1971). Studies on norepinephrine-containing afferents to Purkinje cells of rat cerebellum. I. Localization to the fibers and their synapses, *Brain Res.* **25**:501–521.

Bloom, F. E., G. R. Siggins, and B. J. Hoffer (1974). Interpreting the failures to gigantocellular tegmental field: selectivity of discharge in desynchronized sleep, *Science* **174:**1250–1252.

McCarley, R. W. and J. A. Hobson (1975*a*). Discharge patterns of cat pontine brain stem neurons during desynchronized sleep, *J. Neurophysiol.* **38:**751–766.

McCarley, R. W. and J. A. Hobson (1975*b*). Neuronal excitability modulation over the sleep cycle: a structural and mathematical model, *Science* **189:**58–60.

McCarley, R. W. and J. A. Hobson (1976). PGO waves: phase locked firings by pontine reticular neurons, *Abstr. Annu. Meet. Soc. Neurosci.*, 6th, Toronto, p. 894.

McGinty, D. J. (1973). Neurochemically defined neurons: behavioral correlates of unit activity of serotonin-containing cells, pp. 244–267 in *Brain Unit Activity During Behavior,* Phillips, M. I., ed. Charles C. Thomas, Springfield, Ill.

McGinty, D. M., R. M. Harper, and M. K. Fairbanks (1973). 5-HT-containing neurons: unit activity in behaving cats, pp. 267–279 in *Serotonin and Behavior,* Barchas, J. and E. Usdin, eds. Acadamic Press, New York.

Moruzzi, G. (1972). The sleep-waking cycle, *Ergeb. Physiol.* **64:**1–165.

Mosco, S. S., D. Haubrich, and B. L. Jacobs (1977). Serotonergic afferents to the dorsal raphe nucleus: evidence from HRP and synaptosomal uptake studies, *Brain Res.* **119:**269–290.

Nakamura, S. (1975). Two types of inhibitory effects upon brain stem reticular neurons by low frequency stimulation of raphe nucleus in the rat, *Brain Res.* **93:**140–144.

Nakamura, S. (1977). Some electrophysiological properties of neurons of rat locus coeruleus, *J. Physiol.* **267:**641–658.

Nauta, W. J. H. and H. G. J. M. Kuypers (1958). Some ascending pathways in the brain stem reticular formation, pp. 3–30 in *Reticular Formation of the Brain,* Jasper, H. H. et al., eds. Little, Brown and Co., Boston.

Oliver, A. P. and M. Segal (1974). Transmembrane changes in hippocampal neurons: hyperpolarizing actions of norepinephrine, cyclic AMP, and locus coeruleus, *Progr. Abstr. Annu. Meet. Soc. Neurosci.*, 4th, St. Louis, p. 361.

Olszewski, J. and D. Baxter (1954). *Cytoarchitecture of the Human Brain Stem.* J. B. Lippincott Co., Philadelphia.

Peterson, B. W. (1977). Identification of reticulospinal projections that may participate in gaze control. Paper presented at the Satellite Symposium on Gaze Control, International Congress of Physiology, Paris.

Peterson, B. W., M. Anderson, and M. Filion (1974). Responses of pontomedullary reticular neurons to cortical, tectal, and cutaneous stimuli, *Exp. Brain Res.* **21:**19–44.

Peterson, B. W., R. A. Maunz, N. G. Pitts, and R. G. Mackel (1975). Patterns of projection and branching of reticulospinal neurons, *Exp. Brain Res.* **23:**333–351.

Pickel, V. M., T. H. John, and D. J. Reis (1977). A serotonergic innervation of noradrenergic neurons in nucleus locus coeruleus: demonstration by immunocytochemical localization of the transmitter specific enzymes tyrosine and tryptophan hydroxylase, *Brain Res.* **131:**197–214.

Pivik, R. T., R. W. McCarley, and J. A. Hobson (1977). Eye movement-associated discharge in brain stem neurons during desynchronized sleep, *Brain Res.* **121:**59–76.

Ramón y Cajal, S. (1952). *Histologie du Système Nerveux*, Vol. 1. Consejo Superior de Investigaciones Científicas, Madrid.

Sakai, K., M. Touret, D. Salvert, L. Leger, and M. Jouvet (1977). Afferent projections to the cat locus coeruleus as visualized by the horseradish peroxidase technique, *Brain Res.* **119:**21–41.

Scheibel, M. E. and A. B. Scheibel (1958). Structural substrates for integrative patterns in the brain stem reticular core, pp. 31–55 in *Reticular Formation of the Brain*, Jasper, H. H. et al., eds. Little, Brown and Co., Boston.

Scheibel, M. E. and A. B. Scheibel (1973). Discussion, in *BIS Conference Report No. 32*. Brain Information Service/Brain Research Institute, University of California, Los Angeles.

Segal, M. (1975). Physiological and pharmacological evidence for a serotonergic projection to the hippocampus, *Brain Res.* **94:**115–131.

Segal, M. and F. E. Bloom (1974a). The action of norepinephrine in the rat hippocampus. I. Iontophoretic studies, *Brain Res.* **72:**79–97.

Segal, M. and F. E. Bloom (1974b). The action of norepinephrine in the rat hippocampus. II. Activation of the input pathway, *Brain Res.* **72:**99–114.

Selverston, A. I. and B. Mulloney (1974). Synaptic and structural analysis of a small neural system, pp. 389–395 in *The Neurosciences: Third Study Program*, Schmitt, F. O. and F. G. Worden, eds. MIT Press, Cambridge, Mass.

Siegel, J. M. and D. J. McGinty (1977). Pontine reticular formation neurons: relationship of discharge to motor activity, *Science* **196:**678–680.

Siegel, J. M., D. J. McGinty, and S. M. Breedlove (1977). Sleep and waking activity of pontine gigantocellular field neurons, *Exp. Neurol.* **56:**553–573.

Siggins, G. R., B. J. Hoffer, and F. E. Bloom (1971a). Studies on norepinephrine-containing afferents to Purkinje cells of rat cerebellum. III. Evidence for mediation of norepinephrine effects by cyclic 3′, 5′-adenosine monophosphate, *Brain Res.* **25:**535–553.

Siggins, G. R., B. J. Hoffer, and F. E. Bloom (1971b). Response to Godfraind and Pumain, *Science* **174:**1258–1259.

Siggins, G. R., A. P. Oliver, B. J. Hoffer, and F. E. Bloom (1971). Cyclic adenosine monophosphate and norepinephrine effects on transmembrane properties of cerebellar Purkinje cells, *Science* **171:**192–194.

Sladek, J. R., Jr. (1971). Difference in the distribution of catecholamine varicosities in cat and rat reticular formation, *Science* **174:**410–412.

Steriade, M. and J. A. Hobson (1976). Neuronal activity during the sleep-waking cycle, *Prog. Neurobiol. (Oxford)* **6:**155–376.

Strumwasser, F. (1967). Neurophysiological aspects of rhythms, pp. 516–528 in *The Neurosciences: A Study Program*, Quarton, G. C., T. Melnechuk, and F. O. Schmitt, eds. Rockefeller University Press, New York.

Swanson, L. W. (1976). The locus coeruleus: a cytoarchitectonic Golgi and immunohistochemical study in the albino rat, *Brain Res.* **110:**39–56.

Taber-Pierce, E., W. E. Foote, and J. A. Hobson (1976). The efferent connection of the nucleus raphe dorsalis, *Brain Res.* **107:**137–144.

Torvik, A. and A. Brodal (1957). The origin of reticulospinal fibers in the cat, *Anat. Rec.* **128:**113–135.

Valverde, F. (1961). Reticular formation of the pons and medulla oblongata. A Golgi study, *J. Comp. Neurol.* **116:**71–99.

Van Twyver, H. and T. Allison (1970). Sleep in the opossum *Didelphis marsupialis, Electroencephalogr. Clin. Neurophysiol.* **29:**181–189.

Vertes, R. P. (1977). Selective firing of rat pontine gigantocellular neurons during movement and REM sleep, *Brain Res.* **128:**146–152.

Walberg, F. (1974). Crossed reticulo-reticular projections in the medulla, pons and mesencephalon. An autoradiographic study in the cat, *Z. Anat. Entwicklungsgesch.* **143:**127–134.

Wang, R. Y. and G. K. Aghajanian (1977). Antidromically identified serotonergic neurons in the rat midbrain raphe: evidence for collateral inhibition, *Brain Res.* **132:**186–193.

Wang, R. Y., D. W. Gallager, and G. K. Aghajanian (1976). Stimulation of pontine reticular formation suppresses firing of serotonergic neurons in the dorsal raphe, *Nature (London)* **264:**365–368.

Watanabe, A., S. Obara, and T. Akiyama (1967). Pacemaker potentials for the periodic burst discharge in the heart ganglion of a stomatopod, *Squilla oratoria, J. Gen. Physiol.* **50:**839–862.

Weiss, K. R. and I. Kupfermann (1978). Serotonergic neuronal activity and arousal of feeding behavior in *Aplysia californica*, pp. 66–89 in *Society for Neuroscience Symposia,* Vol. III, Ferrendelli, J. A., ed. Society for Neuroscience, Bethesda, Md.

Winfree, A. T. (1967). Biological rhythms and the behavior of populations of coupled oscillators, *J. Theor. Biol.* **16:**15–42.

Wyzinski, P. W., R. W. McCarley, and J. A. Hobson (1978). Discharge properties of pontine reticulospinal neurons during the sleep-waking cycle, *J. Neurophysiol.,* in press.

NEUROCHEMISTRY OF OLFACTORY CIRCUITS

Chairman's Introduction

The topic of this symposium is novel and merits a comment. One usually thinks of the olfactory system in terms of its chemical sensory function, that of detecting and discriminating among molecules of behavioral significance in the air. From this standpoint it has traditionally attracted the interest of chemists and psychophysicists. This symposium is focused on a different chemical aspect, that of the neurochemistry of the neuronal structures and circuits within the system itself. The development of new techniques in recent years has placed powerful tools at our disposal for analyzing biochemical composition, synthesis and action of neurotransmitters, dynamic changes in metabolism, and many other properties of the nervous system. Among the various regions that have been studied, the olfactory system has turned out to be quite advantageous for the application of many of these methods. The following papers bring together some of this new work for the first time.

The contributions begin with an overview of the anatomical structures and pathways in the olfactory system by Richard Broadwell, emphasizing the extensive interconnections within the system and with other parts of the brain and indicating the evidence for putative transmitters in some of the pathways. Frank Margolis deals with the biochemical constituents of the peripheral olfactory pathway and the evidence for the presence of a specific protein and a dipeptide of unusual interest. Two papers cover recent work on neurotransmitter identification in the olfactory bulb. Charles Ribak describes immunocytochemical identification of GABAergic interneurons and discusses their functional role in dendrodendritic circuits. Stephen Hunt and Jakob Schmidt describe some recent studies on alpha-bungarotoxin binding and the search for cholinergic neurons in the olfactory bulb. The use of the deoxyglucose method to map the functional organization of the olfactory bulb by activity-related glucose uptake is another topic. Finally, Donald Pfaff summarizes work on hormone binding in central olfactory regions and the relation between olfactory and limbic systems in the mediation of reproductive behavior in mammals.

These contributions might have been extended to include other recent studies such as microiontophoretic analysis of transmitters in the olfactory bulb, synaptosomal preparations of dendrodendritic terminals, and localization of dopamine-sensitive adenylate cyclase in the olfactory tubercle. However, the contributions here admirably illustrate the range of present work, from the sensory receptors in the periphery to central systems for control of complex behavior. A most exciting prospect in all of this work is the promise of powerful new insights into the nature of olfactory organization by close correlation of the results with ongoing studies of anatomy and physiology.

NEUROTRANSMITTER PATHWAYS IN THE OLFACTORY SYSTEM

Richard D. Broadwell

National Institute of Neurological and
Communicative Disorders and Stroke, Bethesda, Maryland

INTRODUCTION

The anatomy and connections of the mammalian olfactory system have been subjects of intense investigation with a variety of neuro-anatomical and electrophysiological techniques throughout the past century. Two reasons, among many, for this sustained interest are the role of smell in patterned forms of behavior such as mating and the fact that the olfactory bulb, because of its distinct laminations and several well-differentiated types of neurons, serves as an excellent model of functional organization within the nervous system. The Golgi and experimental silver impregnation methods have been instrumental in providing a fundamental knowledge of the anatomical organization of the olfactory system; however, the recently introduced axoplasmic transport tract tracing methods (i.e., autoradiography, histofluorescence, and horseradish peroxidase) have expanded this understanding in greater detail than was possible with the earlier methods. In the short time the axonal transport methods have been available, they have not only clarified many controversial points concerning the organization of the olfactory system but have demonstrated extensive interrelationships among olfactory-related structures. Among these important findings are the contribution of tufted cells as an additional source of efferent projections from the olfactory bulb (Broadwell and Jacobowitz, 1976; Haberly and Price, 1977; Skeen and Hall, 1977), the identification of a vast plexus of olfactory association projections directed ipsilaterally and contralaterally (Price, 1973; Broadwell, 1975*b*; Broadwell and Jacobowitz, 1976; Haberly and Price, 1978), and the discovery of multiple origins, pathways, and sites of termination of centrifugal fibers to the main and accessory olfactory bulbs (Barber and Field, 1975; Swanson and Hartman, 1975; Broadwell, 1975*b*; Conrad and Pfaff, 1976; Domesick, 1976; Broadwell and Jacobowitz, 1976; Haberly and Price, 1978; Fallon and Moore, 1978*a,b*). With respect to the phylogeny of the mammalian olfactory system, axoplasmic transport studies of the olfactory projections in monkey (Turner, Gupta, and Mishkin, 1978; Rosene, Heimer, Van Hoesen, and Nauta, 1978) suggest that, despite the disparity between primates and rodents on the phylogenetic scale, the connections of the olfactory bulb in the monkey may not be too dissimilar from those in the rat or rabbit.

This presentation will provide an overview of the anatomy and neuronal connections of the mammalian olfactory system. Attention will be given to neurotransmitters believed to act within the olfactory brain, as determined by histofluorescence and enzyme histochemistry, and to the possible involvement of these substances in mediating centrally directed olfactory impulses.

For the purpose of this discussion the olfactory brain is defined as the main and accessory olfactory bulbs, their afferent and efferent pathways, the olfactory cortex, and the centrally located origins of centrifugal fibers to the olfactory bulbs. The olfactory cortex is defined herein as that part of the cortex that receives efferent fibers from the olfactory bulbs. In macrosmatic mammals (i.e., rabbit and rat) the olfactory cortex occupies nearly the entire surface of the cortex ventral to the rhinal fissure (Figure 1a,b). Conversely, in the primate, a supposedly microsmatic mammal, only a small portion of the temporal lobe and basal surface of the cortex are olfactory in function (Turner et al., 1978).

MAIN AND ACCESSORY OLFACTORY BULBS AND THEIR EFFERENT PATHWAYS

The main olfactory bulb consists of seven layers of cells and dendritic and axonal processes concentrically arranged around the rostral extension of the lateral cerebral ventricle (Figure 2a). From the outermost to the innermost lamina are: (1) A layer of olfactory nerve fibers originating from receptor cells located in the olfactory nasal epithelium. (2) The glomerular layer, which includes the olfactory glomeruli, periglomerular and short-axon interneurons, and axon collaterals from tufted cells. Within the glomeruli, synapses are formed

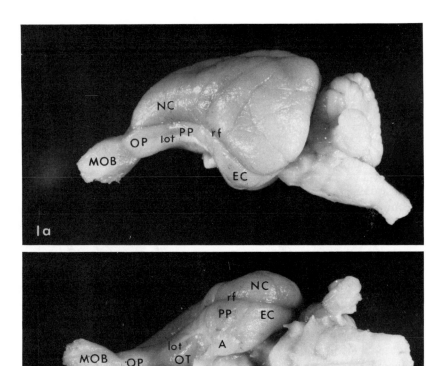

FIGURE 1. Lateral (a) and ventral (b) views of the rabbit brain, illustrating the olfactory cortical structures lying ventral to the rhinal fissure (rf). MOB, main olfactory bulb; OP, olfactory peduncle; OT, olfactory tubercle; lot, lateral olfactory tract; PP, prepiriform cortex; A, amygdala; EC, entorhinal cortex; NC, neocortex.

among terminal arborizations of the incoming olfactory nerve fibers, the primary dendrites of the mitral and tufted cells, and dendritic tufts of the periglomerular neurons (Figure 2b). Periglomerular neurons provide for interglomerular association and are involved in intraglomerular dendrodendritic synaptic connections with mitral and tufted cells. (3)

TERTIARY OLFACTORY PROJECTIONS TO THE LIMBIC SYSTEM AND DIENCEPHALON

Experimental evidence suggests that olfactory input can greatly influence mammalian sexual, mating, and maternal behaviors, the onset of puberty, and hormonal aspects of reproduction (Bruce, 1970; Estes, 1973; Whitten and Champlin, 1973; Powers and Winans, 1975; Winans and Powers, 1977). These complicated behavioral and physiological processes are, in part, under the control of the limbic system and diencephalon, most notably the hypothalamus. Electrical stimulation of the main olfactory bulb activates neurons within the hippocampus (Cragg, 1960; Yokota, Reeves, and MacLean, 1970), hypothalamus (Scott and Leonard, 1971; Scott and Pfaffman, 1972), and the mediodorsal thalamic nucleus (Benjamin and Jackson, 1974; Jackson and Benjamin, 1974). Although the olfactory bulbs do not connect directly with the hippocampal formation or diencephalon, neurons in these structures can be driven through relays in the olfactory cortex. The major neuroanatomical pathways linking the olfactory bulbs with the limbic system and hypothalamus are indicated in Figure 5. The mediodorsal thalamic nucleus receives projections from the olfactory tubercle and prepiriform cortex by way of the stria medullaris and inferior thalamic peduncle (Powell, Cowan, and Raisman, 1965; Scott and Leonard, 1971; Heimer, 1972; Heimer and Wilson, 1975; Skeen, 1976; Siegel, Fukushima, and Edinger, 1976). The mediodorsal nucleus, in turn, projects directly to the orbitofrontal cortex (Leonard, 1969; Krettek and Price, 1977a) and has reciprocal connections with the basolateral amygdaloid nucleus (Krettek and Price, 1977a, b). The relay of olfactory information to the cortex first, the thalamus second, and then back to the cortex is a characteristic unique to the olfactory system and distinguishes it from all other sensory systems.

Autoradiography suggests that the olfactory cortical projection to the mediodorsal nucleus may arise from the paleocortical half of the claustrum or endopiriform nucleus rather than from the prepiriform cortex (Krettek and Price, 1977b). If this is so, then the mediodorsal nucleus may come under the indirect influence of the anterior olfactory nucleus (Broadwell, 1975b).

CENTRIFUGAL FIBERS TO THE OLFACTORY BULBS

Central control or modulation of incoming information from the periphery is common to virtually all sensory systems within the

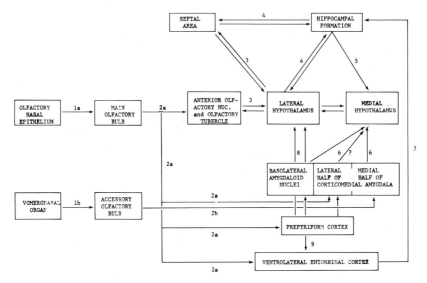

FIGURE 5. Olfactory-related pathways to the hippocampus and hypothalamus. 1a: Olfactory nerves (Land and Shepherd, 1974). 1b: Vomeronasal nerves (Barber and Raisman, 1974). 2a: Lateral olfactory tract (Powell et al., 1965; Heimer, 1968; Price, 1973; Scalia and Winans, 1975; Broadwell, 1975a). 2b: Accessory olfactory tract (Broadwell, 1975a; Scalia and Winans, 1975; Skeen and Hall, 1977). 3: Medial forebrain bundle (Powell et al., 1965; Raisman, 1966; Millhouse, 1969; Scott and Leonard, 1971; Heimer, 1972; Scott and Chafin, 1975; Broadwell, 1975b). 4: Precommissural fornix (Raisman, Cowan, and Powell, 1965; Raisman, 1966; Siegel, Edinger, and Ohgami, 1974; Swanson and Cowan, 1977a, b; Meibach and Siegel, 1977). 5: Postcommissural fornix and medial corticohypothalamic tract (Raisman et al., 1965; Raisman, 1970; Swanson and Cowan, 1977b; Meibach and Siegel, 1977). 6: Stria terminalis (Cowan, Raisman, and Powell, 1965; Heimer and Nauta, 1969; Raisman, 1970; Leonard and Scott, 1971; De Olmos, 1972). 7: Lateral perforant temporo-ammonic tract (Hjorth-Simonsen, 1972; Van Hoesen and Pandya, 1975; Steward, 1976; Steward and Scoville, 1976). 8: Ventral amygdalofugal pathway (Lundberg, 1960; Cowan et al., 1965; Powell et al., 1965; Valverde, 1965; Heimer, 1972; De Olmos, 1972). 9: (Powell et al., 1965; Price, 1973; Krettek and Price, 1977c).

vertebrate nervous system. In many instances this central modulation is expressed through what have come to be called "centrifugal" fiber systems. These fibers originate in higher neural centers and are directed either to relay stations within the ascending sensory pathway or to peripheral receptors. On a priori grounds, the centrifugal fiber system might serve to prevent unwanted or irrelevant signals from reaching higher levels. For example, fibers from each of the sensory areas of the neocortex feed back to related sensory relay nuclei in the thalamus. Fibers from the somatic sensory cortex are also directed to the dorsal column nuclei in the medulla for potential surround inhibition of selected ascending sensory impulses. The olivocochlear bundle in the auditory system, the γ-efferent fibers to muscle spindles, and centrifugal fibers to the retina in the avian visual system are primary examples of a direct central feedback upon a peripheral receptor. The olfactory system, likewise, has its own centrifugal fiber system. These fibers are directed to the main and accessory bulbs and, as will be discussed below, originate from multiple sources within the forebrain and brainstem.

The first detailed description of centrifugal fibers to the main olfactory bulb is credited to Ramón y Cajal (1911, 1955), who distinguished two types: thin anterior commissural fibers and thick ipsilateral fibers. The anterior commissural fibers were thought by Cajal to be of tufted cell origin and to interconnect the two olfactory bulbs. Cajal speculated that the thin fibers were from the ipsilateral sphenoidal (olfactory) cortex of the lateral olfactory tract. Later experimental studies tended to support the commissural connection between the two olfactory bulbs (Fox and Schmitz, 1943; Clark and Meyer, 1947; Fox, Fisher, and Desalva, 1948; Meyer and Allison, 1949; Adey, 1953; Allison, 1953a; Johnson, 1959). Many of Cajal's contemporaries disagreed with him on this subject and proposed that the anterior commissural fibers were derived from more caudal sources, such as the retrobulbar area or olfactory peduncle (Gudden, 1870; Ganser, 1882; Loewenthal, 1897; Van Gehuchten, 1904). Golgi and fiber degeneration investigations in a variety of mammals have since generated the belief that the olfactory peduncle, more commonly known as the anterior olfactory nucleus, is indeed the origin of the commissural projection to the contralateral olfactory bulb and anterior olfactory nucleus (Young, 1941, 1942; Lohman, 1963; Powell et al., 1965; Valverde, 1965; Girgis and Goldby, 1967; Ferrer, 1969; Price and Powell, 1970a). The anterior olfactory nucleus is also believed to send fibers into the ipsilateral olfactory bulb (Valverde, 1965; Ferrer, 1969; Price and Powell, 1970a). One group of investigators suggested that in the rabbit, commissural fibers from the olfactory bulb exist but originate from anterior olfactory nucleus neurons scattered throughout the periventricular layer of the olfactory bulb (Lohman and Mentink, 1969). Autoradiography of the efferent fibers from the olfactory bulb in rat (Price, 1973), rabbit (Broadwell, 1975a), and tree shrew (Skeen and Hall, 1977) rules out commissural projections. Similar results have been obtained with the peroxidase method (Broadwell,

1975*b*; Broadwell and Jacobowitz, 1976; Dennis and Kerr, 1976; Haberly and Price, 1978).

A combined autoradiographic and peroxidase study of the connections of the anterior olfactory nucleus in the rabbit (Broadwell, 1975*b*) indicates this nucleus does project bilaterally to the main olfactory bulb as well as to the external, lateral, and dorsal cell groups of the contralateral anterior olfactory nucleus. Such input to the bulbs is derived predominantly from the pars externa cell group of the anterior olfactory nucleus, with contributions from the pars dorsalis and pars lateralis (Figure 6). In the rat, the pars posterior of the anterior olfactory nucleus is also involved in the contralateral projection to the olfactory bulb (Broadwell and Jacobowitz, 1976). Autoradiograms show that the bilateral projections of the anterior olfactory nucleus into the olfactory bulbs are to the granule cell and glomerular layers (Figure 7). In the granule cell layer, these fibers may terminate upon processes and perikarya of the granule cells and short-axon interneurons. In the glomerular layer, termination may be in relation to periglomerular and short-axon neurons surrounding the olfactory glomeruli. The bilateral projection to the periglomerular region is particularly interesting, as it suggests that the anterior olfactory nucleus can influence olfactory neural activity at the first synaptic relay in the olfactory system.

Based on results with the experimental silver methods, the prepiriform cortex (Cragg, 1962; Heimer, 1968) and olfactory tubercle (Powell et al., 1965) have been implicated as sources for the ipsilateral centrifugal fibers to the main olfactory bulb. These findings were questioned by Price and Powell (1970*b*), who used anterograde fiber and retrograde cell degeneration methods to determine the origins of the centrifugal fibers in the rat. They reported that the only cell group caudal to the anterior olfactory nucleus from which centrifugal fibers appeared to arise was the nucleus of the horizontal limb of the diagonal band. Because fibers from the diagonal band nucleus to the olfactory bulb were traced through the lateral olfactory tract, Price and Powell proposed that the fiber degeneration within the olfactory bulb reported by those who had lesioned the olfactory tubercle or anterior prepiriform cortex was very likely the result of damage to the fibers of passage from the diagonal band. Evoked potential studies, on the other hand, suggest that virtually all structures receiving projections from the olfactory bulb, and not the diagonal band nucleus, contribute to the centrifugal fiber system (Dennis and Kerr, 1968, 1975). The HRP method has been of particular value in settling the ipsilateral centrifugal fiber controversy. Certainly the

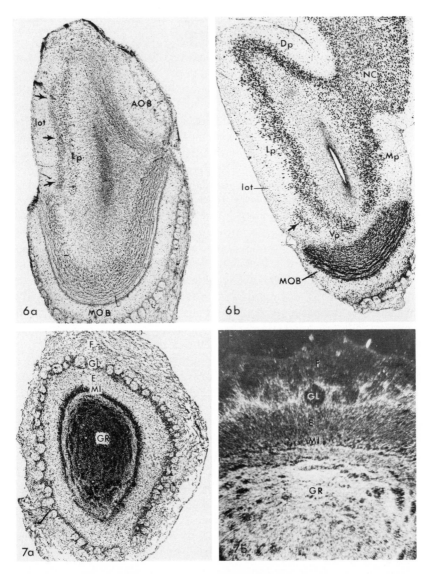

FIGURE 6.   Nissl-stained frontal sections through the intrabulbar (a) and middle (b) portions of the anterior olfactory nucleus in the rabbit to illustrate the cell divisions of the anterior olfactory nucleus. The arrows indicate the pars externa cell group of the anterior olfactory nucleus adjacent to the pars lateralis (Lp). AOB, accessory bulb; MOB, main bulb; Dp, pars dorsalis; Vp, pars ventralis; Mp, pars medialis; NC, neocortex; lot, lateral olfactory tract.

primary attribute of the intra-axonal retrograde transport of peroxidase is the sensitivity of the method in revealing multiple origins of neuronal projections afferent to the site injected with the protein. With this in mind, the HRP technique has shown unequivocally that the ipsilateral centrifugal fibers to the main olfactory bulb are indeed derived from widespread sources within the basal forebrain and also from specific cell groups in the brainstem (see Table 2; Broadwell and Jacobowitz, 1976). Further investigation, with improved HRP methodology, of the origins of the centrifugal fibers to the olfactory bulb in rabbit and rat has been conducted in our laboratory since publication of our initial findings. In addition to numerous perikarya retrogradely labeled with HRP in the nucleus of the horizontal limb of the diagonal band, lateral preoptic area, rostral lateral hypothalamus, and dorsolateral hypothalamus, labeled cell bodies were found in the anterior septal area just ventral to the tenia tecta and in the nucleus of the lateral olfactory tract (Figures 8, 13, 14). A few HRP-labeled somata were scattered among unlabeled cell bodies in the vertical limb of the diagonal band as well. The greatest number of labeled cells found beyond that reported previously was in the prepiriform cortex. Here, HRP-labeled perikarya were plentiful in layers II and III rostral to the anterior amygdaloid area (Figure 9). Within more caudal levels of this cortex, labeled cell bodies were fewer and more widely scattered between layers II and III. This finding of the prepiriform cortex as a prominent source of centrifugal fibers to the olfactory bulb in rabbit and rat parallels similar results obtained with the HRP method in the cat by Dennis and Kerr (1976). Unlike the latter investigators, however, we did not observe HRP-labeled pyramidal cells in the olfactory tubercle, although some labeled cells were scattered deep to layer III of the tubercle. This discrepancy may be attributed to a species variation.

FIGURE 7.   a: Bright-field autoradiogram of a frontal section through the main olfactory bulb ipsilateral to an injection of [³H]leucine into the anterior olfactory nucleus of the rabbit.

b: Dark-field photomicrograph of the autoradiogram in Figure 7a illustrating the projections (white) from the anterior olfactory nucleus to the granule cell (GR) and glomerular (GL) layers of the main olfactory bulb. Similar projections are visible in autoradiograms of the main olfactory bulb contralateral to the injection. With the peroxidase method, these projections have been shown to arise from the pars externa, pars lateralis, and pars dorsalis of the anterior olfactory nucleus. F, olfactory nerve fiber layer; GL, glomerular layer; E, external plexiform layer; Ml, mitral cell layer; GR, granule cell layer.

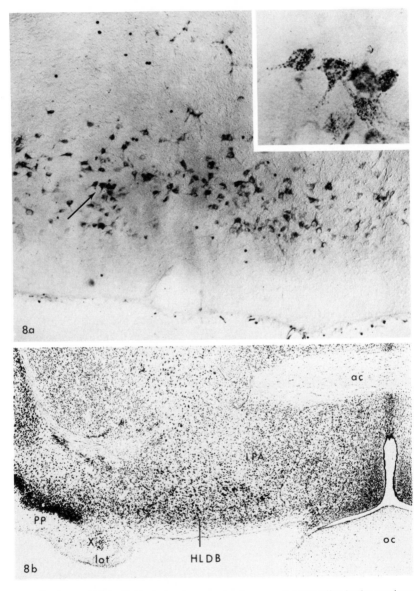

FIGURE 8. a: Bright-field photomicrograph of neuronal cell bodies in the nucleus of the horizontal limb of the diagonal band labeled with peroxidase by retrograde intra-axonal transport following injection of the protein into the ipsilateral main olfactory bulb of the rabbit. The inset (top right) is a higher magnification of the HRP-labeled perikarya indicated by the arrow at the left.

That the olfactory bulb may be influenced by neurons within the brainstem was first suggested on electrophysiological grounds by Ardunini and Moruzzi (1953). Later, with the advent of the histochemical fluorescence method, Dahlström, Fuxe, Olsson, and Ungerstedt (1965) observed noradrenalin- and serotonin-containing fibers and terminals innervating the olfactory bulb. They speculated that these fibers belonged to ascending monoamine neuron systems located in the brainstem. Subsequent histofluorescence and peroxidase histochemistry investigation has proven Dahlström et al. were correct in their interpretation (Broadwell and Jacobowitz, 1976). Our fluorescent rabbit and rat material confirms what Dahlström et al. had reported (Figure 12), and HRP injection into the main olfactory bulb yields retrogradely labeled cell bodies in the locus coeruleus and midbrain raphe (Figures 10, 11, 15). In the rat, HRP-labeled perikarya were located in the locus coeruleus bilaterally. Neurons of the midbrain raphe and the locus coeruleus are known to be serotonergic and noradrenergic, respectively (Dahlström and Fuxe, 1964). Although we did not find HRP-labeled perikarya in any of the dopaminergic cell groups in the brain, Fallon and Moore (1978a, b) have traced efferent projections from the vicinity of the substantia nigra and dopaminergic A10 group in the ventral tegmentum of the mesencephalon to the main olfactory bulb, anterior olfactory nucleus, and prepiriform cortex. The A10-dopaminergic innervation of the olfactory bulb is, in all likelihood, minor in comparison to the noradrenergic-locus coeruleus innervation. The concentration of dopamine within the olfactory bulb is reported to be in precursor amounts of norepinephrine only (Miliaressis, Thoa, Tizabi, and Jacobowitz, 1975).

Aside from the monoamines, acetylcholine is the only other putative neurotransmitter known to be represented within the olfactory centrifugal fiber system. Using acetylcholinesterase histochemistry, Shute and Lewis (1967) traced prominent projections to the glomerular and mitral cell layers of the rat olfactory bulb from acetyl-cholinesterase-staining cell bodies located predominantly in the lateral preoptic area. They referred to this presumptive cholinergic fiber system as the "olfactory radiation." Cell bodies staining intensely for acetylcholinesterase and choline acetyltransferase are heavily concen-

---

b: The same section through the nucleus of the horizontal limb of the diagonal band (HLDB) as seen in Figure 8a, but counterstained with thionin. PP, prepiriform cortex; X, bed nucleus of the accessory olfactory tract; lot, lateral olfactory tract; LPA, lateral preoptic area; ac, anterior commissure; oc, optic chiasm.

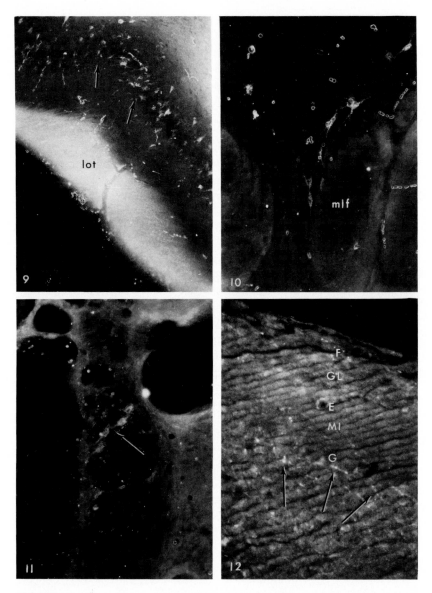

FIGURES 9, 10, and 11. Dark-field photomicrographs of cell bodies in the prepiriform cortex (Figure 9), midbrain raphe (Figure 10), and locus coeruleus (Figure 11) retrogradely labeled with HRP following injection of peroxidase into the ipsilateral main olfactory bulb of the rabbit. Lot, lateral olfactory tract; mlf, medial longitudinal fasciculus.

trated in the diagonal band nucleus, in the lateral preoptic area, and in the rostrolateral and dorsolateral areas of the hypothalamus (Jacobowitz and Palkovits, 1974; Palkovits, Saavedra, Kobayashi, and Brownstein, 1974). These are the same locations in which we find numerous perikarya retrogradely labeled with HRP following injection of peroxidase into the main olfactory bulb (Figures 13, 14). Such evidence would suggest a priori that these HRP-labeled neurons represent the cholinergic centrifugal projection to the olfactory bulb. Doubly staining these cell bodies for acetylcholinesterase and HRP, however, has revealed that only a few of the perikarya which are HRP-positive after olfactory bulb injection are acetylcholinesterase-positive as well (Mesulam, Van Hoesen, and Rosene, 1977; Heimer, personal communication). In view of the dense concentration of acetylcholinesterase-staining fibers and terminals within specific laminae of the olfactory bulb and the absence of similarly staining neuronal cell bodies located therein (Burd, Carson, and Hanker, 1977), the question that remains is whether those few acetylcholinesterase-HRP-positive neurons in the diagonal band and lateral hypothalamus that project to the olfactory bulb have extensive axonal collateralization within the bulb or whether there exists a second central source of cholinergic centrifugal projections, such as from acetylcholinesterase-staining cells in the ventral tegmental area (Palkovits and Jacobowitz, 1974).

Of the forebrain and brainstem sources of centrifugal fibers to the olfactory bulb, the anterior olfactory nucleus may exert the most influence in modulating, by way of the olfactory bulb interneurons, the efflux of mitral and tufted cell impulses destined for the olfactory cortex. Nearly all nuclear groups sending fibers to the olfactory bulb connect en route with the anterior olfactory nucleus, which exchanges connections ipsilaterally, contralaterally, and bilaterally with many of the same nuclear groups of forebrain origin (Figure 16). The anterior olfactory nucleus is the only cell group that receives a massive input from the olfactory bulb and feeds projections back to the olfactory bulb *bilaterally*. In addition, interhemispheric connections exist between the anterior olfactory nucleus and the same nucleus of the opposite side. These two examples of reverberating circuitry influencing the olfactory bulb are referred to as "feedback loops" (Price and Powell, 1970c).

FIGURE 12. A frontal section through the main olfactory bulb of the rabbit prepared for fluorescence microscopy. Fluorescing varicosities (arrows) within the granule cell layer (GR) are indicative of a catecholamine innervation.

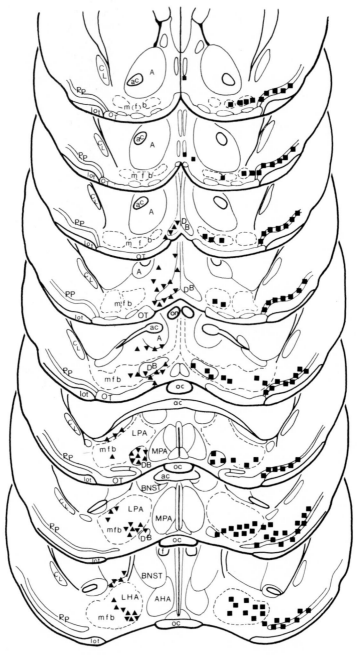

FIGURE 13.   See caption on page 153.

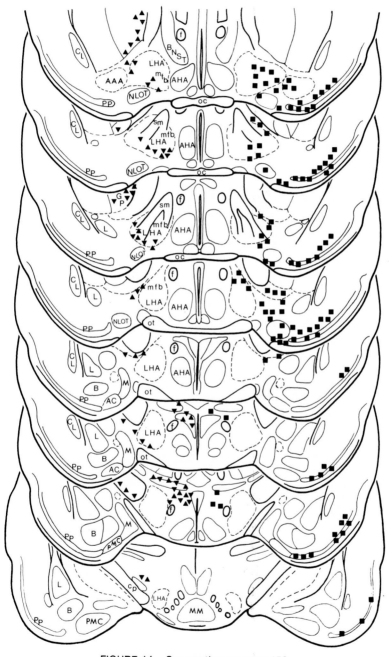

FIGURE 14. See caption on page 153.

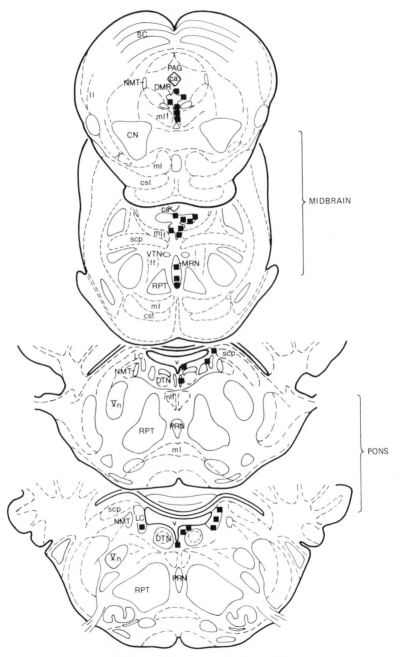

FIGURE 15.   See caption on page 153.

Similar feedback loops exist between the prepiriform cortices bilaterally (Price, 1973) and between the main olfactory bulb and both the prepiriform cortex and the nucleus of the lateral olfactory tract. Feedback loops affecting the activity of the anterior olfactory nucleus, and therefore the olfactory bulb indirectly, are between the anterior olfactory nucleus and the prepiriform cortex bilaterally (Broadwell, 1975b; Broadwell and Jacobowitz, 1976), between the nucleus of the lateral olfactory tract and the prepiriform cortex bilaterally (Haberly and Price, 1978), and possibly between the anterior olfactory nucleus and the

---

FIGURES 13, 14, and 15. Composite diagrams of the distribution of HRP-labeled perikarya in the forebrain and brainstem of the rabbit and rat following injection of the protein into the ipsilateral main olfactory bulb. Concentrations of HRP-positive cell bodies are indicated by squares at the right. Concentrations of acetylcholin-esterase-staining cell bodies (from Jacobowitz and Palkovits, 1974) are indicated by triangles at the left.

Abbreviations

A, Nucleus accumbens
AAA, Anterior amygdaloid area
AC, Anterior cortical amygdaloid nucleus
ac, Anterior commissure
AHA, Anterior hypothalamic area
B, Basal amygdaloid nucleus
BNST, Bed nucleus of the stria terminalis
ca, Cerebral aqueduct
CL, Claustrum
CN, Cuneiform nucleus
cp, Cerebral peduncle
cst, Corticospinal tract
DB, Diagonal band nucleus
DMR, Dorsal midbrain raphe nucleus
DTN, Dorsal tegmental nucleus
f, Fornix
GP, Globus pallidus
L, Lateral amygdaloid nucleus
LC, Locus coeruleus
LHA, Lateral hypothalamic area
ll, Lateral lemniscus
lot, Lateral olfactory tract
LPA, Lateral preoptic area
M, Medial amygdaloid nucleus
mfb, Medial forebrain bundle
ml, Medial lemniscus

mlf, Medial longitudinal fasciculus
MM, Medial mammillary nucleus
MPA, Medial preoptic area
MRN, Medial or central raphe nucleus
NLOT, Nucleus of the lateral olfactory tract
NMT, Nucleus of the mesencephalic tract
oc, Optic chiasm
on, Optic nerve
OT, Olfactory tubercle
ot, Optic tract
PAG, Periaqueductal grey
PMC, Posteromedial cortical nucleus of the amygdala
PP, Prepiriform cortex
PRN, Pontine raphe nucleus
RPT, Reticular nucleus of the pontine tegmentum
SC, Superior colliculus
scp, Superior cerebellar peduncle
sm, Stria medullaris
tt, Tectospinal tract
v, Fourth ventricle
VTN, Ventral tegmental nucleus
Vn, Trigeminal motor nucleus

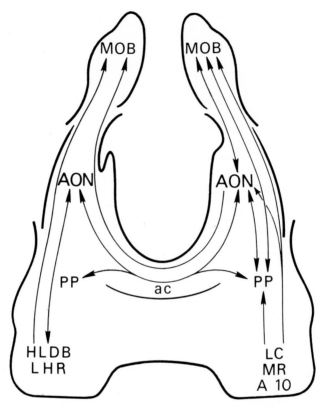

FIGURE 16. Diagram summarizing the neuronal connections among the main olfactory bulb (MOB), anterior olfactory nucleus (AON), and sources of centrifugal fibers to these two structures (see Broadwell, 1975a, b; Broadwell and Jacobowitz, 1976; Haberly and Price, 1978; Fallon and Moore, 1978a, b). For clarity, the reciprocal connections between the olfactory bulb and nucleus of the lateral olfactory tract have been omitted. PP, prepiriform cortex; LC, locus coeruleus; MR, midbrain raphe; A10, dopaminergic cell group in the ventral tegmental area; ac, anterior commissure; LHR, lateral hypothalamic region; HLDB, horizontal limb of the diagonal band.

ipsilateral nucleus of the horizontal limb of the diagonal band (Broadwell, 1975b; Broadwell and Jacobowitz, 1976). The latter cell group can influence the contralateral olfactory bulb and anterior olfactory nucleus through anterior commissural connections with the horizontal limb of the diagonal band nucleus of the opposite side (Price and Powell, 1970c). Because of its afferent and efferent connections, the anterior olfactory nucleus may very well represent a nodal point in the circuits that connect the olfactory bulb with the basal forebrain.

Centrifugal fibers are also directed to the accessory olfactory bulb from multiple sources. Autoradiographic (Barber and Field, 1975; Conrad and Pfaff, 1976) and peroxidase (Broadwell and Jacobowitz, 1976; Haberly and Price, 1978) investigations have shown that the same structures in the forebrain which receive projections from the accessory bulb feed fibers back to the accessory bulb (Table 2). These centrifugal fibers were described initially in degeneration experiments as passing forward in the stria terminalis for termination upon the granule cells within the accessory bulb (Raisman, 1972; De Olmos, 1972). The same monoaminergic cell groups in the brainstem that send projections into the main olfactory bulb also connect with the accessory bulb (Broadwell and Jacobowitz, 1976; Fallon and Moore, 1978a).

FUNCTIONAL ORGANIZATION WITHIN THE OLFACTORY BULB

Termination of commissural and ipsilateral centrifugal fibers within the olfactory bulb is primarily in the granule cell layer and periglomerular region (Cragg, 1962; Powell et al., 1965; Dahlström et al., 1965; Heimer, 1968; Price, 1969; Price and Powell, 1970b; Swanson and Hartman, 1975; Broadwell, 1975b; Broadwell and Jacobowitz, 1976; Fallon and Moore, 1978a, b). At the ultrastructural level, centrifugal axon terminals are presynaptic to granule cell perikarya and processes in the granule cell and external plexiform layers (Price, 1968; Price and Powell, 1970a) and to periglomerular and short-axon cells in the glomerular layer (Pinching and Powell, 1972). Tufted and mitral cell perikarya and dendrites may also receive terminals of the centrifugal fibers (Shute and Lewis, 1967; Pinching and Powell, 1972; Burd et al., 1977).

TABLE 2. *Origins and proposed neurotransmitters (in parentheses) of centrifugal fibers to the main and accessory olfactory bulbs in rabbit and rat*

Main olfactory bulb	Accessory olfactory bulb
Anterior olfactory nucleus bilaterally Anterior septal area Prepiriform cortex Horizontal limb of the nucleus of the diagonal band ⎱ (acetylcholine) Lateral hypothalamic region ⎰ Nucleus of the lateral olfactory tract	Bed nucleus of the accessory olfactory tract Corticomedial complex of the amygdala Bed nucleus of the stria terminalis
└─────── Locus coeruleus (norepinephrine) ───────┐ └─────── Midbrain raphe (serotonin) ───────┘ A10 (dopamine)	

Electrophysiological and biophysical evidence suggests that granule and periglomerular cells are inhibitory interneurons to mitral and tufted cell excitability (Phillips, Powell, and Shepherd, 1962; Shepherd, 1963a, b, 1971; Rall, Shepherd, Reese, and Brightman, 1966; Rall and Shepherd, 1968; Nicoll, 1969; Getchell and Shepherd, 1975). Inhibition of the mitral and tufted cells would be mediated through the reciprocal dendrodendritic synapses between gemmules of the granule cell peripheral processes and the secondary dendrites of mitral and tufted cells in the external plexiform layer (Rall et al., 1966; Price and Powell, 1970a; Willey, 1973) and between dendrites of the periglomerular cells and the mitral and tufted cell dendritic tufts within the glomeruli (Reese and Brightman, 1970; Hinds, 1970; Pinching and Powell, 1971; Reese and Shepherd, 1972; White, 1972, 1973). Excitation of the periglomerular and granule cells by mitral and tufted cells is possible through the reciprocal synapses and recurrent axon collaterals.

Pharmacological (McLennan, 1971; Nicoll, 1971) and immunohistochemical (Ribak, Vaughn, Saito, Barber, and Roberts, 1977) investigations suggest γ-aminobutyric acid is the mediator of granule cell to mitral cell inhibition. Dopamine (Halász, Ljungdahl, Hökfelt, Johansson, Goldstein, Park, and Biberfeld, 1977) and γ-aminobutyric acid (Ribak et al., 1977) have been found within periglomerular neurons. Both substances may be active at the dendrodendritic synapses in the glomeruli. Acetylcholine, serotonin, and noradrenalin administered electrophoretically to the mitral cell layer are reported to decrease the firing rate of mitral cells (Bloom, Costa, and Salmoiraghi, 1964), a response elicited most likely by the action of these agents on the processes and somata of the granule cells. Additional pharmacological data favor an excitatory action for noradrenalin on the granule cells (McLennan, 1971). The ultrastructural studies of Price and Powell (1970a) and Pinching and Powell (1971, 1972) suggest that all centrifugal fibers to the granule, periglomerular, and short-axon cells are excitatory. Synapses with which the centrifugal fibers have been identified as the presynaptic element have asymmetrical membrane thickenings and spheroidal vesicles. These are features believed to be characteristic of excitatory synapses, in contradistinction to symmetrical membrane thickenings and flattened vesicles for inhibitory synapses. As a cautionary note, however, such morphological criteria for identifying presumptive excitatory and inhibitory synapses are equivocal. A given neurotransmitter may exert both excitatory and inhibitory actions, depending upon the type of chemoreceptors in the postsynaptic membrane. Like the periglomerular and granule cells, short-axon cells are thought to be inhibitory interneurons. Axons of short-axon cells synapse with flattened vesicles and symmetrical membrane thickenings onto granule, periglomerular, and other short-axon cells. Additional sources of afferents to short-axon cells are axon collaterals from mitral and tufted cells, centrifugal fibers, and periglomerular cells. The principal connections among the different types of neurons within the olfactory bulb are diagrammed in Figure 17.

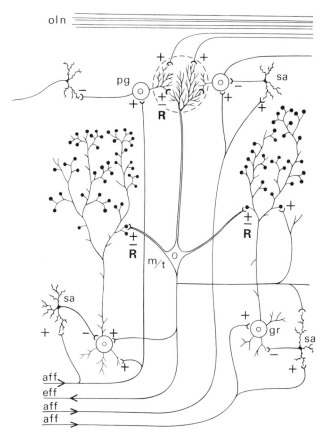

FIGURE 17. Schematic diagram of proposed interconnections among the various types of neurons and the centrifugal fibers within the main olfactory bulb. + or − refers to presumptive excitatory or inhibitory synaptic activity. ±R signifies the reciprocal dendrodendritic synapses within the glomerular and external plexiform layers. See text for discussion and references. Aff, afferent (centrifugal) fibers; eff, efferent fibers from mitral and tufted cells; gr, granule cell; m/t, mitral/tufted cell; pg, periglomerular neuron; sa, short-axon neuron; oln, olfactory nerve fibers. For an interpretation of the neuronal interactions within the olfactory bulb not considered above, see Freeman (1975).

Possible physiological interactions among these neurons are considered in detail elsewhere (Shepherd, 1972; Freeman, 1975).

One final point regarding the centrifugal fibers to the main olfactory bulb deserves brief consideration. Available data suggest that perhaps some degree of preference exists for termination of centrifugal fibers within specific laminae of the olfactory bulb. With histofluorescence the serotonergic centrifugal fibers appear to be directed solely to the glomerular layer, whereas the noradrenergic fibers are directed largely, if not exclusively, to the granule cell layer (Dahlström et al., 1965; Broadwell and Jacobowitz, 1976). The prepiriform cortex also projects predominantly to the granule cell layer (Haberly and Price, 1978). The anterior olfactory nucleus, however, sends fibers to the granule cell and glomerular layers (Broadwell, 1975b). Whatever functional significance, if any, this differential laminar input may have on overall neuronal activity within the olfactory bulb remains to be determined.

MAIN AND ACCESSORY OLFACTORY SYSTEMS

Studies attributing a specific role in reproductive behavior and physiology to the vomeronasal organ and accessory olfactory bulb (Estes, 1973; Powers and Winans, 1975; Winans and Powers, 1977), together with neuroanatomical investigations contrasting the connections of the main and accessory olfactory bulbs (Raisman, 1972; Scalia and Winans, 1975; Broadwell, 1975a; Broadwell and Jacobowitz, 1976; Skeen and Hall, 1977), support the concept of two separate and distinct olfactory systems in macrosmatic mammals. The two olfactory systems might be appropriately referred to as the primary or main olfactory system and the accessory olfactory system. Each differs from the other in its source of sensory input, efferent projections, fiber pathways, and sources of centrifugal fibers (Tables 1 and 2). Both systems may also differ in the efferent projections of the olfactory cortical structures receiving afferent fibers from the olfactory bulbs (Figure 5). Although the efferent projections of the medial half of the corticomedial amygdaloid complex, or that part of the amygdala in receipt of accessory bulb efferents, are well documented, no published information is available on those projections from the lateral half (i.e., anterior cortical and posterolateral cortical amygdaloid nuclei), which receives its input from the main olfactory bulb. The efferent connections of the lateral half of the corticomedial amygdala may be identical to those from the prepiriform cortex or may parallel projections from the medial half of the corticomedial amygdala to the medial hypothalamus. The only apparent overlap of connections between the primary and accessory olfactory systems is provided by the monoaminergic centrifugal fibers. This similarity may be insignificant in view of the fact that the entire cortex, including the olfactory cortex, is likewise innervated by the same monoaminergic cell groups in the brainstem that project to the olfactory bulbs (Fuxe, 1965; Andén, Dahlström, Fuxe, Larsson, Olsson, and Ungerstedt, 1966; Ungerstedt, 1971; Pickel, Segal, and Bloom, 1974; Conrad, Leonard, and Pfaff, 1974; Haberly and Price, 1975; Bobillier, Seguin, Petitjean, Salvert, Tornet, and Jouvet, 1976; Fallon and Moore, 1978a).

REFERENCES

Adey, W. R. (1953). An experimental study of the central olfactory connections in a marsupial (Trichosurus vulpecula), Brain 76:311–330.

Allison, A. C. (1953a). The structure of the olfactory bulb and its relationship to the olfactory pathways in the rabbit and the rat, J. Comp. Neurol. 98:309–355.

Allison, A. C. (1953b). The morphology of the olfactory system in the vertebrates, Biol. Rev. 28:195–244.

Andén, N. E., A. Dahlström, K. Fuxe, K. Larsson, L. Olsson, and U. Ungerstedt (1966). Ascending monoamine neurons to the telencephalon and diencephalon, Acta Physiol. Scand. 67:313–326.

Ardunini, A. and G. Moruzzi (1953). Sensory and thalamic synchronisation in the olfactory bulb, Electroencephalogr. Clin. Neurophysiol. 5:235–242.

Barber, P. C. and P. Field (1975). Autoradiographic demonstration of afferent connections of the accessory olfactory bulb in the mouse, Brain Res. 85:201–203.

Barber, P. C. and G. Raisman (1974). An autoradiographic investigation of the projection of the vomeronasal organ to the accessory olfactory bulb in the mouse, Brain Res. 81:21–30.

Benjamin, R. M. and J. C. Jackson (1974). Unit discharges in the mediodorsal nucleus of the squirrel monkey evoked by electrical stimulation of the olfactory bulb, Brain Res. 75:181–191.

Bloom, F. E., E. Costa, and G. C. Salmoiraghi (1964). Analysis of individual rabbit olfactory bulb neuron responses to the microelectrophoresis of acetylcholine, norepinephrine, and serotonin synergists and antagonists, J. Pharmacol. Exp. Ther. 146:16–23.

Bobillier, P., S. Seguin, F. Petitjean, D. Salvert, M. Tornet, and M. Jouvet (1976). The raphe nuclei of the cat brain stem: a topographical atlas of their efferent projections as revealed by autoradiography, Brain Res. 113:449–486.

Broadwell, R. D. (1975a). Olfactory relationships of the telencephalon and diencephalon in the rabbit. I. An autoradiographic study of the efferent connections of the main and accessory olfactory bulbs, J. Comp. Neurol. 16:329–346.

Broadwell, R. D. (1975b). Olfactory relationships of the telencephalon and diencephalon in the rabbit. II. An autoradiographic and horseradish peroxidase study of the efferent connections of the anterior olfactory nucleus, *J. Comp. Neurol.* **164:**389–410.

Broadwell, R. D. and D. M. Jacobowitz (1976). Olfactory relationships of the telencephalon and diencephalon in the rabbit. III. The ipsilateral centrifugal fibers to the olfactory bulbar and retrobulbar formations, *J. Comp. Neurol.* **170:**321–346.

Bruce, H. M. (1970). Pheromones, *Br. Med. Bull.* **26:**10–13.

Burd, G. D., K. A. Carson, and J. S. Hanker (1977). Light and electron microscopic localization of acetylcholinesterase in the main and accessory olfactory bulbs of the mouse, *Abstr. Annu. Meet. Soc. Neurosci.,* 7th, Anaheim, p. 77.

Clark, W. E. L. and M. Meyer (1947). The terminal connections of the olfactory tract in the rabbit, *Brain* **70:**304–328.

Conrad, L. C., C. M. Leonard, and D. Pfaff (1974). Connections of the medial and dorsal raphe nuclei in the rat. An autoradiographic and degeneration study, *J. Comp. Neurol.* **156:**179–206.

Conrad, L. C. A. and D. W. Pfaff (1976). Efferents from medial basal forebrain and hypothalamus in the rat. I. An autoradiographic study of the medial preoptic area, *J. Comp. Neurol.* **169:**185–220.

Cowan, W. M., G. Raisman, and T. P. S. Powell (1965). The connexions of the amygdala, *J. Neurol. Neurosurg. Psychiatry* **28:**137–151.

Cragg, B. G. (1960). Responses of the hippocampus to stimulation of the olfactory bulb and of various afferent nerves in five mammals, *Exp. Neurol.* **2:**547–571.

Cragg, B. G. (1962). Centrifugal fibres to the retina and olfactory bulb and composition of the supraoptic commissures in the rabbit, *Exp. Neurol.* **5:**406–427.

Dahlström, A. and K. Fuxe (1964). Evidence for the existence of monoamine neurons in the central nervous system. I. Demonstration of monoamines in the cell bodies of brainstem neurons, *Acta Physiol. Scand.* **62** (Suppl. 232):1–80.

Dahlström, A., K. Fuxe, L. Olsson, and U. Ungerstedt (1965). On the distribution and possible function of monoamine nerve terminals in the olfactory bulb of the rabbit, *Life Sci.* **4:**2071–2074.

Dennis, B. J. and D. I. B. Kerr (1968). An evoked potential study of centripetal and centrifugal connections of the olfactory bulb in the cat, *Brain Res.* **11:**373–346.

Dennis, B. J. and D. I. B. Kerr (1975). Olfactory bulb connections with basal rhinencephalon in the ferret: an evoked potential and neuroanatomical study, *J. Comp. Neurol.* **154:**129–148.

Dennis, B. J. and D. I. B. Kerr (1976). Origins of olfactory bulb centrifugal fibres in the cat, *Brain Res.* **110:**593–600.

De Olmos, J. S. (1972). The amygdaloid projection field in the rat as studied with the cupric silver method, pp. 145–204 in *The Neurobiology of the Amygdala,* Eleftherieu, Basil E., ed. Plenum Publishing Corp., New York.

Devor, M. (1976). Fiber trajectories of olfactory bulb efferents in the hamster, *J. Comp. Neurol.* **166:**31–48.

Domesick, V. B. (1976). Projections of the nucleus of the diagonal band of Broca in the rat, *Anat. Rec.* **184:**391–392.

Estes, R. D. (1973). The role of the vomeronasal organ in mammalian reproduction, *Mammalia* **36:**315–341.

Fallon, J. H. and R. Y. Moore (1978a). Catecholamine innervation of the basal forebrain. III. Olfactory bulb, anterior olfactory nuclei, olfactory tubercle, and piriform cortex, *J. Comp. Neurol.*, in press.

Fallon, J. H. and R. Y. Moore (1978b). Catecholamine innervation of the basal forebrain. IV. Topography of the dopamine projection to the basal forebrain and neostriatum, *J. Comp. Neurol.*, in press.

Ferrer, N. G. (1969). Efferent projections of the anterior olfactory nucleus, *J. Comp. Neurol.* **137**:309–320.

Fox, C. A., R. R. Fisher, and S. J. Desalva (1948). The distribution of the anterior commissure in the monkey (*Macaca mulatta*), *J. Comp. Neurol.* **89**:245–277.

Fox, C. A. and J. T. Schmitz (1943). A Marchi study of the distribution of the anterior commissure in the cat. *J. Comp. Neurol.* **79**:297–314.

Freeman, W. J. (1975). *Mass Action in the Nervous System*. Academic Press, New York.

Fuxe, K. (1965). Evidence for the existence of monoamine neurons in the central nervous system. IV. Distribution of monoamine nerve terminals in the central nervous system, *Acta Physiol. Scand.* **64** (Suppl. 247):37–84.

Ganser, S. (1882). Vergleichende anatomische Studien über das Gehirn des Maulwurfs, *Morphol. Jahrb.* **7**:591–725.

Getchell, T. V. and G. M. Shepherd (1975). Short-axon cells in the olfactory bulb: dendrodendritic synaptic interactions, *J. Physiol. (London)* **251**:523–548.

Girgis, M. and F. Goldby (1967). Secondary olfactory connections and the anterior commissure in the coypu, *Myocaster coypus, J. Anat.* **101**:33–44.

Gudden, B. (1870). Experimentaluntersuchungen über das peripherische und zentrale Nervensystem, *Arch. Psychiatr. Nervenkr.* **2**:693–723.

Haberly, L. B. and J. L. Price (1975). Afferent connections of the olfactory cortex, *Abstr. Annu. Meet. Soc. Neurosci.*, 5th, New York, p. 14.

Haberly, L. B. and J. L. Price (1977). The axonal projection patterns of the mitral and tufted cells of the olfactory bulb in the rat, *Brain Res.* **129**:152–157.

Haberly, L. B. and J. L. Price (1978). Association and commissural fiber systems of the olfactory cortex of the rat, *J. Comp. Neurol.*, in press.

Halász, N., A. Ljungdahl, T. Hökfelt, O. Johansson, M. Goldstein, D. Park, and P. Biberfeld (1977). Transmitter histochemistry of the rat olfactory bulb. I. Immunohistochemical localization of monoamine synthesizing enzymes. Support for intrabulbar, periglomerular dopamine neurons, *Brain Res.* **126**:455–474.

Heimer, L. (1968). Synaptic distribution of centripetal and centrifugal nerve fibers in the olfactory system of the rat. An experimental anatomical study, *J. Anat.* **103**:413–432.

Heimer, L. (1972). The olfactory connections of the diencephalon in the rat, *Brain Behav. Evol.* **6**:484–523.

Heimer, L. and W. J. H. Nauta (1969). The hypothalamic distribution of the stria terminalis in the rat, *Brain Res.* **13**:284–297.

Heimer, L. and A. Peters (1968). An electron microscope study of a silver stain for degeneration boutons, *Brain Res.* **8**:337–346.

Heimer, L. and R. D. Wilson (1975). The subcortical projections of the allocortex: similarities in the neural associations of the hippocampus, the piriform cortex, and the neocortex, pp. 177–193 in *Golgi Centennial*

Symposium: Perspectives in Neurobiology, Santini, M., ed. Raven Press, New York.

Hinds, J. W. (1970). Reciprocal and serial dendrodendritic synapses in the glomerular layer of the rat olfactory bulb, *Brain Res.* **17**:530–534.

Hjorth-Simonsen, A. (1972). Projection of the lateral part of the entorhinal area to the hippocampus and fascia dentata, *J. Comp. Neurol.* **146**:219–232.

Jackson, J. C. and R. M. Benjamin (1974). Unit discharges in the mediodorsal nucleus of the rabbit evoked by electrical stimulation of the olfactory bulb, *Brain Res.* **75**:193–201.

Jacobowitz, D. M. and M. Palkovits (1974). Topographic atlas of catecholamine and acetylcholinesterase-containing neurons in the rat brain, *J. Comp. Neurol.* **157**:13–28.

Johnson, R. N. (1959). Studies on the brain of the guinea pig. II. The olfactory tracts and fornix, *J. Comp. Neurol.* **112**:121–140.

Krettek, J. E. and J. L. Price (1977*a*). The cortical projections of the mediodorsal nucleus and adjacent thalamic nuclei in the rat, *J. Comp. Neurol.* **171**:157–192.

Krettek, J. E. and J. L. Price (1977*b*). Projections from the amygdaloid complex to the cerebral cortex and thalamus in the rat and cat, *J. Comp. Neurol.* **172**:687–722.

Krettek, J. E. and J. L. Price (1977*c*). Projections from the amygdaloid complex and adjacent olfactory structures to the entorhinal cortex and to the subiculum in the rat and cat, *J. Comp. Neurol.* **172**:723–752.

Land, L. J. and G. M. Shepherd (1974). Autoradiographic analysis of olfactory receptor projections in the rabbit, *Brain Res.* **70**:506–510.

Leonard, C. M. (1969). The prefrontal cortex of the rat. I. Cortical projection of the mediodorsal nucleus. II. Efferent connections, *Brain Res.* **12**:321–343.

Leonard, C. M. and J. W. Scott (1971). Origin and distribution of the amygdalofugal pathways in the rat. An experimental neuroanatomical study, *J. Comp. Neurol.* **141**:313–330.

Loewenthal, S. (1897). Über das Riechhirn der Saugetiere. Festschr. 69. Vers. Dtsch. Naturforsch. u. Ärzte, Braunschweig, *Deitr. Wiss. Med.* 213–220.

Lohman, A. H. M. (1963). The anterior olfactory lobe of the guinea pig. A descriptive and experimental anatomical study, *Acta Anat.* **53** (Suppl. 49):1–109.

Lohman, A. H. M. and G. M. Mentink (1969). The lateral olfactory tract, the anterior commissure and the cells of the olfactory bulb, *Brain Res.* **12**:396–413.

Lundberg, P. O. (1960). Cortico-hypothalamic connexions in the rabbit. An experimental neuroanatomical study, *Acta Physiol. Scand.* **49** (Suppl. 171).

McCotter, R. E. (1912). The connexions of the vomeronasal nerves with the accessory olfactory bulb in the opossum and other mammals, *Anat. Rec.* **6**:299–318.

McLennan, H. (1971). The pharmacology of inhibition of mitral cells in the olfactory bulb, *Brain Res.* **24**:177–184.

Meibach, R. C. and A. Siegel (1977). Efferent connections of the hippocampal formation in the rat, *Brain Res.* **124**:197–224.

Mesulam, M.-M., G. W. Van Hoesen, and D. L. Rosene (1977). Substantia innominata, septal area and nuclei of the diagonal band in the rhesus monkey:

organization of efferents and their acetylcholinesterase histochemistry, *Abstr. Annu. Meet. Soc. Neurosci.*, 7th, Anaheim, p. 202.

Meyer, M. and A. C. Allison (1949). An experimental investigation of the olfactory tracts in the monkey, *J. Neurol. Neurosurg. Psychiatry* **12:**274–286.

Miliaressis, E., N. B. Thoa, Y. Tizabi, and D. M. Jacobowitz (1975). Catecholamine concentration in discrete brain areas following self-stimulation in the ventromedial tegmentum of the rat, *Brain Res.* **100:**142–147.

Millhouse, O. E. (1969). A Golgi study of the descending medial forebrain bundle, *Brain Res.* **15:**341–363.

Nicoll, R. A. (1969). Inhibitory mechanisms in the rabbit olfactory bulb: dendrodendritic mechanisms, *Brain Res.* **14:**157–172.

Nicoll, R. A. (1971). Pharmacological evidence for GABA as the transmitter in granule cell inhibition in the olfactory bulb, *Brain Res.* **35:**137–149.

Palkovits, M. and D. M. Jacobowitz (1974). Topographic atlas of catecholamine and acetylcholinesterase-containing neurons in the rat brain. II. Hindbrain (mesencephalon, rhombencephalon), *J. Comp. Neurol.* **157:**29–42.

Palkovits, M., J. M. Saavedra, R. M. Kobayashi, and M. B. Brownstein (1974). Choline acetyltransferase content of limbic nuclei of the rat, *Brain Res.* **79:**443–450.

Phillips, C. G., T. P. S. Powell, and G. M. Shepherd (1962). Responses of mitral cells to stimulation of the lateral olfactory tract in the rabbit, *J. Physiol. (London)* **168:**65–88.

Pickel, V. M., M. Segal, and F. E. Bloom (1974). A radioautographic study of the efferent pathways of the nucleus locus coeruleus, *J. Comp. Neurol.* **155:**15–42.

Pinching, A. J. and T. P. S. Powell (1971). The neuropil of the glomeruli of the olfactory bulb, *J. Cell Sci.* **9:**347–377.

Pinching, A. J. and T. P. S. Powell (1972). The termination of centrifugal fibers in the glomerular layer of the olfactory bulb, *J. Cell Sci.* **10:**621–635.

Powell, T. P. S., W. M. Cowan, and G. Raisman (1965). The central olfactory connections, *J. Anat.* **99:**791–813.

Powers, J. B. and S. Winans (1975). Vomeronasal organ: critical role in mediating sexual behavior of the male hamster, *Science* **187:**961–963.

Price, J. L. (1968). The termination of centrifugal fibers in the olfactory bulb, *Brain Res.* **7:**483–486.

Price, J. L. (1969). The origin of the centrifugal fibers to the olfactory bulb, *Brain Res.* **14:**542–545.

Price, J. L. (1973). An autoradiographic study of complementary laminar patterns of termination of afferent fibers to the olfactory cortex, *J. Comp. Neurol.* **150:**87–108.

Price, J. L. and T. P. S. Powell (1970*a*). An electron microscopic study of the termination of the afferent fibres to the olfactory bulb from the cerebral hemisphere, *J. Cell Sci.* **7:**157–187.

Price, J. L. and T. P. S. Powell (1970*b*). An experimental study of the origin and the course of the centrifugal fibres to the olfactory bulb in the rat, *J. Anat.* **107:**215–237.

Price, J. L. and T. P. S. Powell (1970*c*). The afferent connections of the nucleus of the horizontal limb of the diagonal band, *J. Anat.* **107:**239–256.

Raisman, G. (1966). The connexions of the septum, *Brain* **89:**317–348.

Raisman, G. (1970). An evaluation of the basic pattern of connections between the limbic system and the hypothalamus, *Am. J. Anat.* **129:**197–202.

Raisman, G. (1972). An experimental study of the projection of the amygdala to the accessory olfactory bulb and its relationship to the concept of a dual olfactory system, *Exp. Brain Res.* **14:**395–408.

Raisman, G., W. M. Cowan, and T. P. S. Powell (1965). An experimental analysis of the efferent projection of the hippocampus, *Brain* **89:**83–108.

Rall, W. and G. M. Shepherd (1968). Theoretical reconstruction of field potential and dendrodendritic synaptic interactions in olfactory bulb, *J. Neurophysiol.* **31:**884–915.

Rall, W., G. M. Shepherd, T. S. Reese, and M. W. Brightman (1966). Dendrodendritic synaptic pathway for inhibition in the olfactory bulb, *Exp. Neurol.* **14:**44–56.

Ramón y Cajal, S. (1911). *Histologie du Système Nerveux de l'Homme et des Vertébrés,* Tome II. A. Maloine, Paris.

Ramón y Cajal, S. (1955). *Studies on the Cerebral Cortex,* Kraft, L., trans. Lloyd-Luke, London.

Reese, T. S. and M. W. Brightman (1970). Olfactory surface and central olfactory connections in some vertebrates, pp. 115–149 in *Ciba Foundation Symposium: Taste and Smell in Vertebrates*, Wolstenholme, G. E. W. and J. Knight, eds. Churchill, London.

Reese, T. S. and G. M. Shepherd (1972). Dendrodendritic synapses in the central nervous system, pp. 121–136 in *Structure and Function of Synapses,* Pappas, G. D. and D. P. Purpura, eds. Raven Press, New York.

Ribak, C., J. E. Vaughn, K. Saito, R. Barber, and E. Roberts (1977). Glutamate decarboxylase localization in neurons of the olfactory bulb, *Brain Res.* **126:**1–18.

Rosene, D. L., L. Heimer, G. W. Van Hoesen, and W. J. H. Nauta (1978). The centripetal and centrifugal connections of the olfactory bulb in the rhesus monkey, in preparation.

Scalia, F. and S. Winans (1975). The differential projections of the olfactory bulb in mammals, *J. Comp. Neurol.* **161:**31–55.

Scott, J. W. and B. R. Chafin (1975). Origin of olfactory projections to lateral hypothalamus and nuclei gemini of the rat, *Brain Res.* **88:**64–68.

Scott, J. W. and C. M. Leonard (1971). The olfactory connections of the lateral hypothalamus in the rat, mouse and hamster, *J. Comp. Neurol.* **141:**331–344.

Scott, J. W. and C. Pfaffmann (1972). Characteristics of responses of lateral hypothalamic neurons to stimulation of the olfactory system, *Brain Res.* **48:**251–264.

Shepherd, G. M. (1963a). Responses of mitral cells to olfactory nerve volleys in the rabbit, *J. Physiol. (London)* **168:**89–100.

Shepherd, G. M. (1963b). Neuronal systems controlling mitral cell excitability, *J. Physiol. (London)* **168:**101–117.

Shepherd, G. M. (1971). Physiological evidence for dendrodendritic synaptic interactions in the rabbit's olfactory glomerulus, *Brain Res.* **32:**212–217.

Shepherd, G. M. (1972). Synaptic organization of the mammalian olfactory bulb, *Physiol. Rev.* **52:**864–917.

Shute, C. C. D. and P. R. Lewis (1967). The ascending cholinergic reticular system: neocortical olfactory and subcortical projections, *Brain* **90:**497–522.

Horinishi and Margolis, 1977). We have observed carnosine synthetase activity in the olfactory bulb and mucosa of several mammalian species (unpublished observations), but mouse appears to have the highest specific activity of the various species studied. Carnosinase activity was also present in both olfactory mucosa and bulb, although the former was much more active than the latter (Table 3). Although the difference in tissue levels of carnosinase activity presented here is about 30-fold between olfactory bulb and mucosa, this ratio can vary from 5- to 80-fold depending upon the conditions of the assay (Margolis, Brown, Harding, and Grillo, manuscript in preparation). Carnosinase is present in the olfactory epithelium of species as diverse from the mouse as the salamander *Ambystoma tigrinum* (Margolis and Getchell, unpublished observation). Carnosinase activity exceeds that of carnosine synthetase in all neural tissues studied. In order to evaluate the cellular location of these enzymes in the olfactory tissues, we determined the effects of both peripheral deafferentation by $ZnSO_4$ irrigation and central denervation by bulbectomy on their respective activities. The levels of carnosine and carnosine synthetase declined rapidly in the bulb after deafferentation, concomitant with the loss in marker protein (Table 4). Various other enzyme activities changed little over the same time interval, demonstrating the specificity of the phenomenon and the probable location of the carnosine synthetase activity in the olfactory receptor neuron synapses. In the reciprocal experiment (Table 5), carnosine content, carnosine synthetase activity, and olfactory marker protein content in the mucosa declined rapidly after bulbectomy, indicating their probable common location in the degenerating mature olfactory receptor neurons. Several other enzyme activities, including that of carnosinase, changed little over the course of this experiment, suggesting the proba-

TABLE 4. *Specificity of marker decline in bulb after deafferentation*

Days	Carn	Carn Synthase	OMP	GAD	β-ala-T	ACE	CAT	mg Tissue
0	100	100	100	100	100	100	100	100
7	10	15	27		96			80
30	5	6	17	110	78	81	110	57
60	5	5	10		78			50

Data are presented as % control value at 0 days.
Abbreviations: Carn, carnosine; Carn Synthase, carnosine synthetase; OMP, olfactory marker protein; GAD, glutamic decarboxylate; β-ala-T, β-alanine transaminase; ACE, acetylcholinesterase; CAT, choline acetyltransferase.
Deafferentation by $ZnSO_4$ as in Table 2.

TABLE 5. *Specificity of marker decline in mucosa after bulbectomy*

Days	Carn	Carn Synthase	Carn-ase	OMP	β-ala-T	ACE
0	100	100	100	100	100	100
3	45	46		42		
14	8	18	85	35	80	100

Data are presented as % control value at 0 days.
Abbreviations as for Table 4; Carn-ase, carnosinase.
GAD and CAT are barely detectable in this tissue.

ble extraneuronal location of these enzymes. These data, demonstrating the probable neuronal location of carnosine synthetase and extraneuronal location of carnosinase, fulfill the second of the biochemical criteria of Table 1.

High-affinity uptake was initially thought to reflect neuronal processes exclusively (Kuhar, 1973; Snyder, Yamamura, Pert, Logan, and Bennett, 1973). Since the demonstration that glia also contribute to this phenomenon, its significance as an indicator of putative neurotransmitters has become unclear (Henn, 1975). It is nevertheless of interest to note that peripheral deafferentation is without effect on the high-affinity uptake by the olfactory bulb of a large number of compounds, including the carnosine precursors β-alanine and histidine (Table 6), although the latter two are taken up in vivo (Neidle and Margolis,

TABLE 6. *Effect of peripheral deafferentation on putative neurotransmitter uptake systems in the olfactory bulbs*

Compound	Tissue: medium ratio		Exogenous transmitter (nM)
	Control	3 weeks after $ZnSO_4$	
β-Alanine	0.5 ± 0.1	0.6 ± 0.1	6.0
γ-Aminobutyric acid	24.0 ± 1.4	30.0 ± 3.6	52.7
Choline	2.9 ± 0.2	3.2 ± 0.4	159.0
Dopamine	4.0 ± 0.3	5.2 ± 0.3	65.0
Histidine	4.2 ± 0.5	3.8 ± 0.4	4.0
Glutamic acid	16.0 ± 1.9	15.0 ± 1.2	58.6
Glycine	0.9 ± 0.2	0.9 ± 0.4	42.3
Norepinephrine	1.6 ± 0.3	1.7 ± 0.8	36.7

K_m values for each compound were determined using 6–8 concentrations each, with quadruplicate determinations at each concentration. Lines were computer-constructed by a linear regression program and routinely gave correlation coefficients of 0.95 or better. Data are presented as mean values ± S.D. for analyses performed in at least quadruplicate. Deafferentation by $ZnSO_4$ as in Table 2.

1976) and in vitro (Neidle and Kandera, 1974) and converted to carnosine. No uptake of carnosine per se could be detected in olfactory
bulb at 8 μM exogenous concentration (Keller and Margolis, unpublished observation). Uptake of carnosine by brain tissue has, however,
been reported under similar conditions (Abraham, Pisano, and Uden-
friend, 1964). The reason for this discrepancy is unclear. However,
when we direct our attention to precursor uptake in vivo by the olfactory
mucosa, we see a rather active process. Following irrigation of the
mucosa with radiolabeled β-alanine or histidine, these two precursors
of carnosine are rapidly taken up and converted to carnosine, some
of which is transported to the olfactory bulb (Margolis and Grillo,
1977). The time course for [β-^{14}C]alanine uptake by mucosa, conversion to carnosine, and transport to bulb is seen in Figure 4. This phenomenon is blocked by pretreatment in vivo with $ZnSO_4$, N-ethylmalei-
mide or vinblastine sulfate (Margolis and Grillo, 1977; Harding,
Getchell, and Margolis, 1978), or by sectioning the olfactory nerve
(Harding, Donlan, Chen, and Wright, 1977a; Rochel and Margolis,

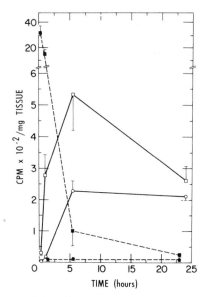

FIGURE 4.   Time course of [1-β-^{14}C]alanine uptake and conversion to carnosine following intranasal administration of 1 μCi per mouse. Data are presented as mean
cpm/mg tissue ± S.D. for n = 4. Carnosine in bulb (○) and mucosa (□); β-alanine in
bulb (●) and mucosa (■).

unpublished observations). Cell suspensions prepared from rat olfactory mucosa also will take up the labeled precursors with formation of carnosine (Hirsch and Margolis, 1977). The olfactory mucosa of the mouse is capable of exhibiting a massive increase in β-alanine content subsequent to administration of β-alanine in vivo (Table 7). This is followed by an increase in tissue carnosine content (Table 7). Thus, evidence for a precursor uptake derives from several directions to satisfy the third criterion of Table 1.

It has proven to be extremely difficult to obtain data with regard to criterion four (Table 1) because of the facility with which carnosine leaks out of tissue during synaptosomal and cellular preparations. This, of course, is not a unique problem with this dipeptide, and methods to overcome the problem are under investigation.

If a compound is to be considered a candidate neurotransmitter, evidence must be adduced in support of a membrane-associated binding site at the tissue location where the compound is postulated to function as a transmitter. Our initial studies of this question utilized the technique of proton magnetic resonance spectroscopy ([1]HMR). Study of the alterations of the [1]HMR spectra of carnosine in the presence of membranes from olfactory mucosa has indicated the presence of a carnosine binding site which is not stereospecific but does require a particular steric configuration of the histidinyl portion of the peptide (Brown, Margolis, Williams, Pitcher, and Elgar, 1977). Attempts to demonstrate a similar site in a membrane fraction from olfactory bulb by [1]HMR techniques were equivocal. Olfactory bulb membranes do bind carnosine, but in a nonsterically-specfic manner (Brown et al., 1977). More recent [1]HMR data (Brown and Margolis, unpublished observation) on carnosine binding by olfactory mucosal membranes show that the binding is unaffected by bulbectomy, indicating it is not associated with

TABLE 7.   *Effect of β-alanine on carnosine in vivo*

Hours after β-alanine	β-Alanine (nmol/mg)			Carnosine (nmol/mg)	
	Bulb	Mucosa	Plasma	Bulb	Mucosa
0	0.15	0.23	N.D.	1.92 ± 0.24	1.83 ± 0.60
2	1.0	10.4			
6	0.66	6.07	0.80	1.79 ± 0.52	2.33 ± 0.69
27	0.19	0.61	N.D.	2.54 ± 0.62	3.06 ± 0.50

β-Alanine was administered at 2 g/Kg i.p.; amino-oxyacetic acid (15 mg/Kg i.p.) was given 15 min before β-alanine. β-Alanine data are average of two assays of two to three mice per assay. Carnosine data are mean ± S.D. for five individual mice. N.D., not detectable.

mature olfactory neurons but must reside in another cell type in the mucosa. To study carnosine binding further, we switched to the more sensitive radioligand receptor assay. While the sensitivity of the magnetic resonance technique was only about 10 nmol/mg dry weight (Brown et al., 1977), that of the radioligand binding assay can be as little as 10 fmol/mg protein, given a ligand of high enough specific activity. This latter method does, however, lack the ability to describe the exact molecular configuration of the bound ligand or its residence time on the binding site, which can, in contrast, be obtained from ^{1}HMR.

Because of the sensitivity of the radioligand binding method, we next attempted to evaluate carnosine binding by olfactory tissue using this method. For this purpose, ^{3}H-labeled L-carnosine was enzymatically synthesized at specific activities in excess of 30 Ci/mmol. Using this material and a minor modification of the equilibrium filter binding assay, we could show L-carnosine binding to a preparation of olfactory bulb membranes with a K_d of about 1 μM and a B_{max} of 300–500 fmol/mg membrane preparation (Hirsch and Margolis, 1978a). In contrast to the nonstereospecific sites detected in the olfactory epithelium and bulb by proton magnetic resonance, this reaction is saturable and stereospecific (Figure 5). The apparent discrepancy between the two techniques is probably related to the existence of more than one site and the relative sensitivities of the ^{1}HMR and radioligand binding techniques. The inability of a large number of drugs, amino acids, and peptides to displace L-[^{3}H]carnosine (Table 8) testifies to the specificity of this L-[^{3}H]carnosine binding site in olfactory bulb membranes. This binding site is trypsin-sensitive, demonstrating its protein nature, and is present at 3- to 15-fold higher levels in olfactory bulb than in several other cerebral tissues tested (Table 9). It is curious to note that mouse muscle, which also contains significant amounts of carnosine (Margolis, 1974a), seems devoid of both this high-affinity stereospecific binding site (Table 9) and the low-affinity nonstereospecific site studied by ^{1}HMR (Brown et al., 1977).

Peripheral deafferentation of the olfactory bulb reduced the stereospecific binding of L-[^{3}H]carnosine to olfactory bulb membranes by >90% in a few days (Figure 6). No return was seen for several months. The binding of a variety of other labeled ligands by bulb membranes declined less than 30% after several months, and for GABA and etorphine was still indistinguishable from control values even 6 months after treatment, when the bulb has atrophied to one-half of its normal weight (Figure 6).

The demonstration of a binding site for L-[^{3}H]carnosine in the olfactory

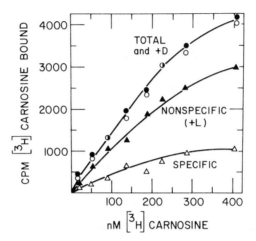

FIGURE 5. Stereospecificity and saturability of L-[³H]carnosine binding to mouse olfactory bulb membranes. Filter binding assays were performed in triplicate at the indicated concentrations of L-[³H]carnosine. Total binding (●) and that in the presence of 1 mM unlabeled D-carnosine (○) are virtually the same. Fewer counts remain bound in the presence of unlabeled 1 mM L-carnosine (▲) and are defined as nonspecific binding. Stereospecific binding (△) was calculated by difference.

bulb, presumably located in the synaptic glomerulus, completes the fifth biochemical criterion for a transmitter candidate listed in Table 1. We have shown that L-[³H]carnosine binding is (1) saturable, (2) reversible, (3) of high affinity and feasible at physiological conditions, (4) stereospecific, (5) restricted in tissue and cellular distribution, and (6) specific with regard to drugs and analogues capable of displacement. Thus our various data on L-[³H]carnosine binding to olfactory bulb membrane preparations fulfill six of the seven criteria (Cuatrecasas and Hollenberg, 1976) which must be satisfied in order to identify a

TABLE 8. *Specificity of displacement of L-[³H]carnosine from bulb membranes*

The following compounds produced no displacement at 0.1 mM: GABA, L-histidine, β-alanine, glycine, L-aspartic acid, L-glutamic acid, L-proline, taurine, histamine, phentolamine, (−)propranolol, haloperidol, chlorpromazine, pyrilamine maleate, diphenhydramine, cimetidine, metiamide, ergothioneine, QNB, dopamine, serotonin, (+)bicuculline, Na pentobarbital, flurazepam, nipecotic acid, amino-oxyacetic acid, pargyline, glycyl-histamine.

The following peptides were essentially ineffective at 0.1 or 1.0 mM: D-carnosine, L-homocarnosine, β-alanyl-β-alanine, β-alanyl-L-alanine, β-alanyl-glycine, glycylglycylglycylglycine, pyroglutamyl-L-histidine or its O-methyl ester.

Baradi, A. F. and G. H. Bourne (1959). Histochemical localization of cho-
linesterase in gustatory and olfactory epithelia, *J. Histochem. Cytochem.*
7:2–7.
Bitensky, M. W., N. Miki, J. J. Keirns, M. Keirns, J. M. Baraban, J. Freeman,
M. A. Wheller, J. Lacy, and F. R. Marcus (1975). Activation of photoreceptor
disk membrane phosphodiesterase by light and ATP, pp. 213–240 in *Advances in Cyclic Nucleotide Research*, Vol. 5, Drummond, G. I., P. Greengard, and G. A. Robison, eds. Raven Press, New York.
Broadwell, R. D. (1978). Neurotransmitter pathways in the olfactory system,
pp. 131–166 in *Society for Neuroscience Symposia*, Vol. III, Ferrendelli,
J. A., ed. Society for Neuroscience, Bethesda, Md.
Broadwell, R. D. and D. M. Jacobowitz (1976). Olfactory relationships of the
telencephalon and diencephalon in the rabbit. III. The ipsilateral centrifugal
fibers to the olfactory bulbar and retrobulbar formations, *J. Comp. Neurol.*
170:321–346.
Brown, C. E., F. L. Margolis, T. H. Williams, R. G. Pitcher, and G. Elgar
(1977). Carnosine in olfaction. Proton magnetic resonance spectral evidence
for tissue-specific carnosine binding sites, *Neurochem. Res.* **2:**555–580.
Cheal, M. L. and R. L. Sprott (1971). Social olfaction: a review of the role
of olfaction in a variety of animal behaviors, *Psychol. Rep.* **29:**195–243.
Chung-Hwang, E., H. Khurana, and H. Fisher (1976). The effect of dietary
histidine level on the carnosine content of rat olfactory bulbs, *J. Neurochem.*
26:1087–1091.
Cuatrecasas, P. and M. D. Hollenberg (1976). Membrane receptors and hormone action, *Adv. Protein Chem.* **30:**251–451.
Dahlström, A., K. Fuxe, L. Olsson, and U. Ungerstedt (1965). On the distribution and possible function of monoamine nerve terminals in the olfactory
bulb of the rabbit, *Life Sci.* **4:**2071–2074.
Doty, R. L. (1976). *Mammalian Olfaction, Reproductive Processes and Behavior.* Academic Press, New York.
Ferriero, D. and F. L. Margolis (1975). Denervation in the primary olfactory
pathway of mice. II. Effects on carnosine and other amine compounds,
Brain Res. **86:**75–86.
Fesenko, E. E., V. I. Novoselov, G. Ya. Pervukhin, and N. K. Fesenko
(1977). Molecular mechanisms of odor sensing. II. Studies of fractions from
olfactory tissue scrapings capable of sensitizing artificial lipid membranes
to action of odorants, *Biochim. Biophys. Acta* **466:**347–356.
Filogamo, G. and P. C. Marchisio (1971). Acetylcholine system and neural
development, *Neurosci. Res.* **4:**29–64.
Gennings, J. N., D. B. Gower, and L. H. Bannister (1977). Studies on the
receptors to 5α-androst-16-en-3-one and 5α-androst-16-en-3α-ol in sow
nasal mucosa, *Biochim. Biophys. Acta* **496:**547–556.
Getchell, T. V. and M. L. Getchell (1974). Signal detecting mechanisms in
the olfactory epithelium: molecular discrimination, in *Odors: Evaluation,
Utilization and Control*, Cain, W. S., ed. *Ann. N. Y. Acad. Sci.* **237:**62–76.
Godfrey, D. A., C. D. Ross, and A. D. Williams (1977). Choline acetyltrans-
ferase and acetylcholinesterase in the olfactory system of the rat, *Abstr.
Annu. Meet. Soc. Neurosci.*, 7th, Anaheim, p. 78.
Goren, E., A. H. Hirsch, and O. M. Rosen (1974). Activity stain for the

detection of cyclic nucleotide phosphodiesterase in polyacrylamide gels, *Methods Enzymol.* **38:**259–261.

Graham, L. T. (1977). Glutamate and aspartate associated with lateral olfactory tract fibers, *Trans. Am. Soc. Neurochem.* **8:** 209.

Graziadei, P. P. C. (1973). Cell dynamics in the olfactory mucosa, *Tissue Cell* **5:**113–131.

Graziadei, P. P. C. and G. A. Monti Graziadei (1978). Continuous nerve cell renewal in the olfactory system, in *Handbook of Sensory Physiology*, Jacobson, M., ed. Springer-Verlag, Berlin, in press.

Gross, G. W. and D. G. Weiss (1977). Subcellular fractionation of rapidly transported axonal material in olfactory nerve: evidence for a size dependent molecule separation during transport, *Neurosci. Lett.* **5:**15–20.

Halász, N., A. Ljungdahl, T. Hökfelt, O. Johansson, M. Goldstein, D. Park, and P. Biberfeld (1977). Transmitter histochemistry at the rat olfactory bulb. I. Immunohistochemical localization of monoamine synthesizing enzymes. Support for intrabulbar, periglomerular dopamine neurons, *Brain Res.* **126:** 455–474.

Harding, J. W., K. Donlan, N. Chen, and J. Wright (1977*a*). Behavioral and biochemical support of olfactory neuron replacement, *Abstr. Annu. Meet. Soc. Neurosci.*, 7th, Anaheim, p. 79.

Harding, J. W., T. V. Getchell, and F. L. Margolis (1978). Denervation of the primary olfactory pathway in mice. V. Long-term effect of intranasal $ZnSO_4$ irrigation on behavior, biochemistry and morphology, *Brain Res.* **140:** 271–285.

Harding, J., P. P. C. Graziadei, G. A. Monti Graziadei, and F. L. Margolis (1977*b*). Denervation in the primary olfactory pathway of mice. IV. Biochemical and morphological evidence for neuronal replacement following nerve section, *Brain Res.* **132:**11–28.

Harding, J. and F. L. Margolis (1976). Denervation in the primary olfactory pathway of mice. III. Effect on enzymes of carnosine metabolism, *Brain Res.***110:**351–360.

Hartman, B. K. and F. L. Margolis (1975). Immunofluroescence localization of the olfactory marker protein, *Brain Res.* **96:**176–180.

Henn, F. A. (1975). Glial transport of amino acid neurotransmitter candidates, pp. 91–97 in *Metabolic Compartmentation and Neurotransmission*, Berl, S., D. D. Clarke, and D. Schneider, eds. Plenum Press, New York.

Hirsch, J. D. and F. L. Margolis (1977). Preparation and characterization of cells from rat olfactory mucosa, *Abstr. Annu. Meet. Soc. Neurosci.*, 7th, Anaheim, p. 79.

Hirsch, J. D. and F. L. Margolis (1978*a*). Carnosine and neurotransmitter binding sites in olfactory bulb membranes, *Trans. Am. Soc. Neurochem.* **9:**149.

Hirsch, J. D. and F. L. Margolis (1978*b*). Cell suspensions from rat olfactory neuroepithelium: biochemical and histochemical characterization, *Brain Res.*, in press.

Horinishi, H. and F. L. Margolis. (1977). Carnosine synthetase from mouse olfactory bulb, *Fed. Proc.* **36:**2357.

Kalyankar, G. and A. Meister (1959). Enzymatic synthesis of carnosine and related β-alanyl and γ-aminobutyryl peptides, *J. Biol. Chem.* **234:**3210–3218.

Kaplan, M. S. and J. W. Hinds (1977). Neurogenesis in the adult rat: electron microscopic analysis of light radioautographs, *Science* **197**:1092–1094.

Keller, A. and F. L. Margolis (1975). Immunological studies of the rat olfactory marker protein, *J. Neurochem.* **24**:1101–1106.

Keller, A. and F. L. Margolis (1976). Isolation and characterization of rat olfactory marker protein, *J. Biol. Chem.* **251**:6232–6237.

Koch, R. B. and D. Desiah (1975). Preliminary studies on rat olfactory tissue, *Life Sci.* **15**:1005–1015.

Kuhar, M. J. (1973). Neurotransmitter uptake: a tool in identifying neuro-transmitter-specific pathways, *Life Sci.* **13**:1623–1634.

Kurihara, K. and N. Koyama (1972). High activity of adenyl cyclase in olfactory and gustatory organs, *Biochem. Biophys. Res. Commun.* **48**:30–34.

Lolley, R. N., D. B. Farber, M. E. Rayborn, and J. G. Hollyfield (1977). Cyclic GMP accumulation causes degeneration of photoreceptor cells: simulation of an inherited disease, *Science* **196**:664–666.

Margolis, F. L. (1974). Carnosine in the primary olfactory pathway, *Science* **184**:909–911.

Margolis, F. L. (1975). Biochemical markers of the primary olfactory pathway: a model neural system, pp. 193–246 in *Advances in Neurochemistry*, Vol. 1, Agranoff, B. W. and M. H. Aprison, eds. Plenum Press, New York.

Margolis, F. L. and M. Grillo (1977). Axoplasmic transport of carnosine (β-alanyl-L-histidine) in the mouse olfactory pathway, *Neurochem. Res.* **2**:507–519.

Margolis, F. L., N. Roberts, D. Ferriero, and J. Feldman (1974). Denervation in the primary olfactory pathway of mice: biochemical and morphological effects, *Brain Res.* **81**:469–483.

Menevse, A., G. Dodd, and T. M. Poynder (1977). Evidence for the specific involvement of cyclic AMP in the olfactory transduction mechanism, *Biochem. Biophys. Res. Commun.* **77**:671–677.

Monti Graziadei, G. A., F. L. Margolis, J. W. Harding, and P. P. C. Graziadei (1977). Immunocytochemistry of the olfactory marker protein, *J. Histochem. Cytochem.* **25**:1311–1316.

Moulton, D. G. (1974). Cell renewal in the olfactory epithelium of the mouse, in *Odors: Evaluation, Utilization and Control*, Cain, W. S., ed. *Ann. N. Y. Acad. Sci.* **237**:52–61.

Neidle, A. and J. Kandera (1974). Carnosine: an olfactory bulb peptide, *Brain Res.* **80**:359–364.

Neidle, A. and F. L. Margolis (1976). Carnosine turnover in olfactory bulb and other tissues of the mouse, *Trans. Am. Soc. Neurochem.* **7**:117.

Orr, M. D., R. L. Blakeley, and D. Panagou (1972). Discontinuous buffer systems for analytical and preparative electrophoresis of enzymes on polyacrylamide gel, *Anal. Biochem.* **45**:68–85.

Price, S. (1973). Phosphodiesterase in tongue: activation by bitter taste stimuli, *Nature (London)* **241**:54–55.

Ribak, C. E., J. E. Vaughn, K. Saito, R. Barber, and E. Roberts (1977). Glutamate decarboxylase localization in neurons of the olfactory bulb, *Brain Res.* **126**:1–18.

Rosenberg, A. (1960). The activation of carnosinase by divalent metal ions, *Biochim. Biophys. Acta* **45**:297–316.

Roskoski, R., L. D. Ryan, and F. J. P. Diecke (1974). γ-Aminobutyric acid synthesized in the olfactory nerve, *Nature (London)* **251**:526–529.

SenGupta, P. (1967). Olfactory receptor reaction to the lesion of the olfactory bulb, pp. 193–201 in *Olfaction and Taste II*, Hayashi, T., ed. Pergamon Press, New York.

Shepherd, G. M. (1970). The olfactory bulb as a simple cortical system: experimental analysis and functional implications, pp. 539–551 in *The Neurosciences: Second Study Program*, Schmitt, F. O., ed. Rockefeller University Press, New York.

Shepherd, G. M. (1972). Synaptic organization of the mammalian olfactory bulb, *Physiol. Rev.* **52**:864–917.

Shepherd, G. M. (1974). *The Synaptic Organization of the Brain*. University Press, London.

Shibuya, T., Y. Aihara, and K. Tonosaki (1977). Single cell responses to odors in the reptilian olfactory bulb, pp. 23–32 in *Food Intake and Chemical Senses*, Katsuki, Y., M. Sato, S. Takagi, and Y. Oomura, eds. University of Tokyo Press, Tokyo.

Snyder, S. H., H. I. Yamamura, C. B. Pert, W. J. Logan, and J. P. Bennett (1973). Neuronal uptake of neurotransmitters and their precursors. Studies with "transmitter" amino acids and choline, pp. 195–222 in *New Concepts of Neurotransmitter Regulation*, Mandell, A., ed. Plenum Press, New York.

Stenesh, J. J. and T. Winnick (1960). Carnosine-anserine from muscle. 4. Partial purification of the enzyme and further studies of β-alanyl peptide synthesis, *Biochim. Biophys. Acta* **77**:575–581.

Takagi, S. F. (1971). Degeneration and regeneration of the olfactory epithelium, pp. 76–94 in *Handbook of Sensory Physiology*, Beidler, L. M., ed. Springer-Verlag, Berlin.

Wells, J. N. and J. G. Hardman (1977). Cyclic nucleotide phosphodiesterases, pp. 119–143 in *Advances in Cyclic Nucleotide Research*, Vol. 8, Greengard, P. and G. A. Robison, eds. Raven Press, New York.

Werman, R. (1966). Criteria for identification of a central nervous system transmitter, *Comp. Biochem. Physiol.* **18**:745–766.

Wideman, J., L. Brink, and S. Stein (1978). New automated fluorometric peptide assay at picomole level: carnosine in olfactory bulb, *Anal. Biochem.*, in press.

THE GABAergic NEURONS OF THE OLFACTORY BULB IN THE RAT

Charles E. Ribak

City of Hope National Medical Center, Duarte, California

INTRODUCTION

Glutamate decarboxylase (GAD), the synthesizing enzyme of the neurotransmitter γ-aminobutyric acid (GABA), has been shown to exhibit different activities within the various layers of the olfactory bulb (Graham, 1973). High levels of GAD activity parallel high concentrations of GABA in the external plexiform layer (EPL), the glomerular layer (GL), and the granule cell layer (GRL). Physiological recordings from mitral cells following stimulation of either the olfactory nerves or the lateral olfactory tract have indicated that the granule and periglomerular cells inhibit the mitral/tufted cell dendrites located in the EPL and GL (Getchell and Shepherd, 1975; Nicoll, 1971; Phillips, Powell, and Shepherd, 1963; Rall, Shepherd, Reese, and Brightman, 1966; Shepherd, 1972). In addition, the GABA antagonists picrotoxin and bicuculline block the inhibitory effect produced by the granule and periglomerular cells (McLennan, 1971; Nicoll, 1971). Thus, the data from these studies suggest that both the granule and periglomerular cells of the olfactory bulb are inhibitory interneurons which use GABA as their neurotransmitter.

In this paper on the olfactory bulb, I will present previously published results obtained with my colleagues at the City of Hope National Medical Center using an immunocytochemical method for the localization of GAD (Ribak, Vaughn, Saito, Barber, and Roberts, 1977). The method used was similar to that already published in previous work (Barber and Saito, 1976; McLaughlin, Barber, Saito, Roberts, and Wu, 1975; McLaughlin, Wood, Saito, Barber, Vaughn, Roberts, and Wu, 1974; Ribak, Vaughn, Saito, Barber, and Roberts, 1976; Wood, McLaughlin,

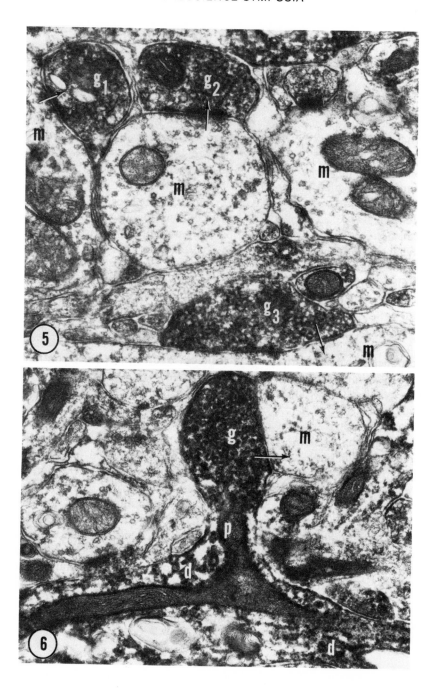

Electron Microscopic Analysis of the Glomerular Layer (GL)

The GL is organized into cellular and neuropil compartments which have been referred to as the periglomerular and the glomerular regions, respectively (Pinching and Powell, 1971a). Pinching and Powell (1971 a, b, c) and White (1973) have described the constituents of these regions, and their results provide the framework for the present analysis. The periglomerular region contains the somata of external tufted cells, periglomerular (PGL) cells, and the superficial short-axon cells. The dendrites of the tufted cells and the PGL cells leave the proximity of their somata to enter into the glomerular region, where they make synaptic contacts with the olfactory nerve endings and with each other. Short-axon cells limit their dendrites to the periglomerular region. The neuropil of the glomeruli contains: (1) the terminal tufts of dendrites from both mitral and tufted cells, (2) the olfactory nerve axon terminals, and (3) the dendrites and gemmules from PGL cells. The dendrites of mitral and tufted cells are grouped together into one category because of their similarities in morphology and connectivity. The dendritic shafts and gemmules of PGL cells enter into reciprocal synaptic relationships with mitral/tufted dendrites and are also involved in serial synaptic relationships.

The somata and dendrites of many PGL cells in the periglomerular region were stained with GAD-positive reaction product in thin sections from specimens incubated in anti-GAD serum (e.g., Figure 7). The distribution of the reaction product within the somata of PGL cells was similar to that found within the somata of the granule cells, i.e., the highest concentration of reaction product occurred around the cisternae and vesicles of the Golgi apparatus. The GAD-positive PGL cells also showed reaction product in their proximal dendrites which entered the glomerular neuropil. The somata and dendrites of the other neuronal

---

FIGURES 5 and 6. Electron micrographs of olfactory bulb incubated in anti-GAD serum. Figure 5 shows GAD-positive gemmules (g_1–g_3) forming synapses (arrows) with mitral cell dendrites (m) in the EPL. Two of these gemmules (g_1 and g_2) are postsynaptic to mitral cell dendrites, while the other GAD-positive gemmule (g_3) seems to be presynaptic (polarity indicated by direction of arrows). ×35,000. In Figure 6, GAD-positive reaction product is concentrated in a granule cell gemmule (g) in the EPL and is also distributed within its dendrite (d) and pedicle (p). This dendrite was identified as a peripheral dendrite of a granule cell because of its radial orientation to the surface of the olfactory bulb. The gemmule of this GAD-positive granule cell dendrite forms a synapse (arrow) with a mitral cell dendrite (m). ×38,000.

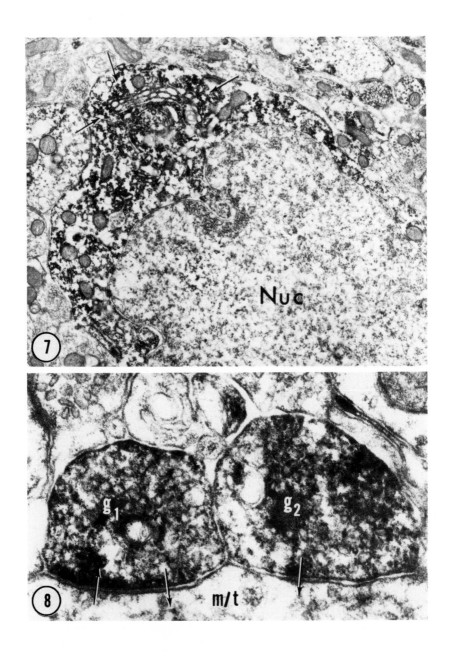

types in the periglomerular region (i.e., short-axon and external tufted neurons) were observed to be free of reaction product.

In the glomerular region, profiles filled with GAD-positive reaction product were analyzed using random and serial thin sections. These GAD-positive profiles were identified as either gemmules or dendritic shafts of PGL cells on the basis of their synaptic relationships and with the use of serial section reconstructions. For example, in Figure 8 a GAD-positive profile is shown forming a reciprocal synapse with a mitral/tufted dendritic shaft. This GAD-positive profile was identified as a PGL gemmule because of its characteristic synaptology (Pinching and Powell, 1971*b*; White, 1973). Other GAD-positive gemmules were reconstructed from serial sections, and some of them were observed to be in continuity with dendritic shafts. Such dendritic shafts also contained GAD-positive reaction product and participated in reciprocal synaptic relationships with mitral/tufted dendrites in the glomerular region. In addition, these GAD-positive dendrites (e.g., Figures 9 and 10) participated in serial synaptic relationships, and thus these dendrites were identified as arising from PGL cells (Pinching and Powell, 1971*b*; White, 1973).

DISCUSSION

The localization of GAD within granule and periglomerular neurons is consistent with the biochemical analysis of GAD activity in the rat olfactory bulb which suggests that both of these cell types use GABA as their neurotransmitter (Graham, 1973). In addition, these results are consistent with the results of physiological and pharmacological studies which indicate that granule and PGL cells inhibit mitral/tufted cells and that this inhibition is mediated by GABA (Getchell and Shepherd, 1975; McLennan, 1971; Nicoll, 1971; Phillips et al., 1963; Rall et al., 1966; Shepherd, 1972). Thus, the data from these studies,

---

FIGURES 7 and 8. Electron micrographs of the glomerular layer of olfactory bulb incubated in anti-GAD serum. Figure 7 shows GAD-positive reaction product throughout the somata of a periglomerular cell, but not within its nucleus (Nuc). The somal reaction product is most concentrated around the Golgi complex (arrows). ×14,000. Figure 8 shows two GAD-positive gemmules (g_1 and g_2) of PGL cells in a glomerulus. One of the gemmules (g_1) appears to form a reciprocal synapse (arrows) with a mitral/tufted dendritic shaft (m/t). The other gemmule (g_2) appears to be presynaptic (arrow) to the same dendritic shaft. The polarities of the synaptic junctions are indicated by the direction of the arrows. ×70,000.

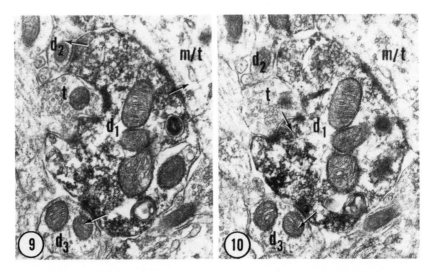

FIGURES 9 and 10.   Electron micrographs from serial sections of a glomerulus in olfactory bulb incubated in anti-GAD serum. These micrographs show a serial synaptic arrangement involving a periglomerular cell dendrite. Figure 9 shows a periglomerular cell dendritic shaft (d₁) containing synaptic vesicles, mitochondria, and GAD-positive reaction product. This dendrite (d₁) is presynaptic to a mitral/tufted dendritic shaft (m/t) and to two other dendritic profiles of unknown origin (d₂ and d₃). The arrows indicate the polarity of each synapse. Figure 10 is an adjacent section showing the same GAD-positive periglomerular cell dendritic shaft (d₁) as in Figure 9. In this section, the GAD-positive dendrite (d₁) appears to be postsynaptic to an axon terminal (t) that is probably derived from a centrifugal fiber (Price and Powell, 1970c). Also, d₁ is shown to be presynaptic to the same small dendritic profile (d₃) as in Figure 9. This periglomerular cell dendrite forms part of a serial synaptic arrangement, since it is presynaptic to three dendritic profiles and is postsynaptic to both an axon terminal and, as determined in another part of the series, the mitral/tufted dendrite (m/t). ×30,000.

in combination with the findings of the immunocytochemical localization of GAD to granule and PGL cells, establish that these cells in the olfactory bulb are inhibitory interneurons which use GABA as their neurotransmitter.

The results of immunocytochemical studies for the localization of other neurotransmitter-synthesizing enzymes indicate that other neurotransmitters are present within the rat olfactory bulb (Halász, Ljungdahl, Hökfelt, Johansson, Goldstein, Park, and Biberfeld, 1977; Hökfelt, Halász, Ljungdahl, Johansson, Goldstein, and Park, 1975). Halász et al. (1977) demonstrate an ultrastructural localization of tyrosine hydroxylase within the dendrites and somata of tufted cells in the external plexiform and glomerular layers, and these results indicate that tufted cells use dopamine as a neurotransmitter, since dopamine-β-hydroxylase was absent from this cell type. Both Hökfelt et al. (1975) and Halász et al. (1977) suggest that PGL cells also contain tyrosine hydroxylase in the glomerular layer, although these studies do not demonstrate tyrosine hydroxylase localization within the gemmules of PGL cells in the glomeruli. Therefore, it remains debatable whether or not some PGL cells use dopamine as a neurotransmitter. The immunocytochemical localization of dopamine-β-hydroxylase and histochemical results in the olfactory bulb suggest that noradrenaline fibers are present within the granule and external plexiform layers and that serotonin fibers are present within the glomerular layer (Dahlström, Fuxe, Olsson, and Ungerstedt, 1965; Halász et al., 1977; Swanson and Hartman, 1975). The neurotransmitter of mitral cells is presently unknown.

The localization of GAD to somata and dendrites in the olfactory bulb represented the first occasion when GAD was not localized exclusively to a neuron's axon terminals. Previous investigations using the same antisera to GAD which were used in the present study had shown that the somata which gave rise to certain GAD-positive axon terminals lacked GAD-positive reaction product (McLaughlin et al., 1974, 1975; Wood et al., 1976). Wood et al. (1976) have recently discussed reasons for this failure to observe GAD in somata. One explanation was based on the fact that crude mitochondrial fractions of the brain, mainly composed of synaptosomes and mitochondria, were used for the biochemical isolation of GAD (Wu, 1976). On this basis it was suggested that GAD in synaptic terminals might exist in an antigenic form which is somewhat different from that located in nonsynaptic neuronal regions. Thus, antisera to synaptosomal GAD might not cross-react with GAD in the rest of the neuron. An alternative explanation by Wood et al. (1976) suggested that somal GAD is present in low, undetectable concentrations relative to GAD in axon terminals because it might be rapidly transported from its site of synthesis to its sites of utilization. Our observation of GAD-positive reaction product in the cell bodies and dendrites of granule and PGL neurons demonstrates that this GAD is in an antigenic form that is very similar to the GAD in synaptic terminals. Therefore, our results support the latter suggestion that current immunocytochemical methods are not sufficiently sensitive to detect small quantities of freshly synthesized GAD in the cell bodies of certain neurons which transport the enzyme to axon terminals.

A possible reason for the detection of GAD in the somata and dendrites of granule and PGL neurons may relate to the fact that the presynaptic sites of these neurons are primarily dendritic. In contrast, the neurons observed in previous GAD immunocytochemical studies (McLaughlin et al., 1974, 1975; Wood et al., 1976) exclusively have axon terminals as their presynaptic sites. In addition to the olfactory bulb, we have observed GAD-positive somata and dendrites in the lateral geniculate body, the medial geniculate body, and the superior colliculus (Ribak et al., 1977). Since these regions of the central nervous system are also known to contain neurons with presynaptic dendrites (Famiglietti, 1970; Lieberman, 1973; Lund, 1969; Morest, 1971; Tigges and Tigges, 1975), this finding is consistent with the possibility that somal and dendritic GAD may be detected by current immunocytochemical methods only within neurons which have presynaptic dendrites.

The fact that GAD can be detected in the somata of neurons which transport this enzyme into presynaptic dendrites indicates that such neurons have a higher somal concentration of GAD than neurons which do not have presynaptic dendrites. This variation in somal concentration suggests that the transport of GAD from the somata of presynaptic dendrite (PSD) neurons (e.g., granule and PGL cells) is different from that of presynaptic axon (PSA) neurons (e.g., Purkinje, Golgi II, basket, and stellate cells of the cerebellum). This concentration difference might also be due to variations in size between PSD and PSA neurons, but this possibility appears unlikely since some cells from the two categories are comparable in size (e.g., granule and stellate cells). Thus, some factors that may contribute to altered transport of GAD are differences in: (1) rates of synthesis of GAD in the cell body, (2) rates of loading GAD onto axonal and dendritic transport mechanisms, (3) velocities of the two transport systems (Schubert and Kreutzberg, 1975), and (4) rates of GAD turnover at presynaptic sites.

Although the exact mechanisms which might cause alterations in the centrifugal flow of GAD in the two neuronal types (PSD and PSA neurons) remain unresolved, a difference in the transport from the somata into their processes would seem to be a reasonable explanation for the apparent variation in GAD concentrations of certain PSD and PSA cells. This possibility has been tested recently by using colchicine to block axonal flow in PSA cells in order to see whether GAD would accumulate to detectable levels within their somata. These experiments have shown that GAD can be detected within the

Schubert, P. and G. W. Kreutzberg (1975). Parameters of dendritic transport, pp. 255–268 in *Advances in Neurology,* Vol. 12, Kreutzberg, G. W., ed. Raven Press, New York.

Shepherd, G. M. (1972). Synaptic organization of the mammalian olfactory bulb, *Physiol. Rev.* **52:**864–917.

Swanson, L. W. and B. K. Hartman (1975). The central adrenergic system. An immunofluorescence study of the location of cell bodies and their efferent connections in the rat utilizing dopamine-β-hydroxylase as a marker, *J. Comp. Neurol.* **163:**467–506.

Tigges, M. and J. Tigges (1975). Presynaptic dendrite cells and two other classes of neurons in the superficial layers of the superior colliculus of the chimpanzee, *Cell Tissue Res.* **162:**279–297.

White, E. L. (1973). Synaptic organization of the mammalian olfactory glomerulus: new findings including an intraspecific variation, *Brain Res.* **60:**299–313.

Wood, J. G., B. J. McLaughlin, and J. E. Vaughn (1976). Immunocytochemical localization of GAD in electron microscopic preparations of rodent CNS, pp. 133–148 in *GABA in Nervous System Function,* Roberts, E., T. N. Chase, and D. B. Tower, eds. Raven Press, New York.

Wu, J.-Y. (1976). Purification, characterization, and kinetic studies of GAD and GABA-T from mouse brain, pp. 7–55 in *GABA in Nervous System Function,* Roberts, E., T. N. Chase, and D. B. Tower, eds. Raven Press, New York.

ARE MITRAL CELLS CHOLINERGIC?

Stephen Hunt and Jakob Schmidt

State University of New York, Stony Brook, New York

INTRODUCTION

Over the past few years the anatomical distribution and molecular properties of receptors for several neurotransmitters have been analyzed by means of high-affinity ligands (Fewtrell, 1976). Among these, α-bungarotoxin (αBuTX), a major component of the venom of *Bungarus multicinctus*, has been used as a specific marker for the nicotinic acetylcholine receptor (AChR) in skeletal muscle (Sytkowski, Vogel, and Nirenberg, 1973; Fertuck and Salpeter, 1974; Barnard, Dolly, Porter, and Albuquerque, 1975; Devreotes and Fambrough, 1975; Lentz, Mazurkiewicz, and Rosenthal, 1977).

Nicotinic AChR are also known to occur in autonomic ganglia and in the central nervous system. This has led to the expectation that, despite pharmacological differences between nicotinic receptors on neural and non-neural cells (Koelle, 1970), αBuTX and related snake toxins may be useful as cholinergic probes in interneuronal synapses.

Indeed, it has been established in recent years that binding sites for αBuTX exist in the autonomic and central nervous system of the rat (Eterović and Bennett, 1974; Salvaterra, Mahler, and Moore, 1975; Polz-Tejera, Schmidt, and Karten, 1975) and of other species besides rodents. It has furthermore been observed that these neuronal αBuTX receptors resemble peripheral nicotinic AChR biochemically (Lowy, MacGregor, Rosenstone, and Schmidt, 1976), that they bind nicotinic and not muscarinic drugs with high affinity (Schmidt, 1977), and that they are synaptic membrane constituents (Bartfai, Berg, Schultzberg, and Heilbronn, 1976; Lentz and Chester, 1977; Hunt and Schmidt, 1978*a*). From a survey of toxin receptor distribution within and outside of the central nervous system, a good correlation between toxin binding activity and expected levels of nicotinic AChR has emerged (Schechter, Handy, Pezzementi, and Schmidt, 1978). A particularly striking example for the concordance of toxin and established cholinergic markers is the rat cerebellum, where choline acetyltransferase, acetylcholinesterase, and αBuTX receptors are found predominantly within folia IX and X (Kása and Silver, 1969; Hunt and Schmidt, 1978b).

Physiological evidence for the specificity of the toxin is largely lacking. Attempts to block synaptic transmission in sympathetic ganglia (Brown and Fumagalli, 1977) or the spinal cord (Miledi and Szczepaniak, 1975; Duggan, Hall, and Lee, 1976) have not been successful, and immunological data indicate that toxin receptors and AChR are distinct membrane components in a cell line derived from adrenal medulla (Patrick and Stallcup, 1977). On the other hand, αBuTX has been reported to block synaptic transmission in the optic tectum of *Bufo marinus* (Freeman, 1977). Considering all available information, one may regard as viable the working hypothesis that αBuTX is a cholinergic marker in the central nervous system, even if the membrane component with which it interacts may serve roles other than that of a classical neurotransmitter receptor.

The present report describes the use of αBuTX as a probe in the olfactory bulb. The toxin binding analysis indicates that the mitral cell may be cholinergic, a proposition tested and supported by an experiment on acetylcholinesterase synthesis. These results are discussed in light of other evidence concerning the chemistry of synaptic transmission in the olfactory bulb.

RESULTS AND DISCUSSION

Toxin

The distribution of αBuTX binding sites within the rat olfactory bulb was studied both in fresh cryostat sections incubated in 10^{-10} M radio-iodinated toxin and in material prepared for autoradiography, 24 hours after intraventricular injection of $[^{125}I]$-αBuTX (for description of methods, see Polz-Tejera et al., 1975; Hunt and Schmidt, 1978a, b). The results of both methods were similar, and the autoradiographic patterns could be abolished by 10^{-3} M D-tubocurarine or nicotine or a 50- to 100-fold excess of native toxin, but not by atropine (a muscarinic cholinergic ligand), attesting to the nicotinic specificity of the binding. With the light microscope, toxin binding sites were found predominantly within the glomeruli and the inner portion of the external plexiform layer (EPL), while background levels of silver grains were seen over the periglomerular area and granule cell layer (Figure 1). The resolution of the light microscopic autoradiographic method was inadequate for an identification of the structures to which the toxin was binding. However, previous studies had indicated that αBuTX interacts with synaptic sites within the central nervous system, which in turn suggested (by exclusion) that, within the olfactory bulb, mitral/tufted cells are the neuronal population most likely to be releasing acetylcholine. The reasoning ran as follows: There are three major cell types within the olfactory bulb: mitral/tufted cells, periglomerular cells, and granule cells (Ramón y Cajal, 1911). Immunohistochemical studies have indicated that the specific synthesizing enzyme for the neurotransmitter GABA is found within the majority of granule cells, while

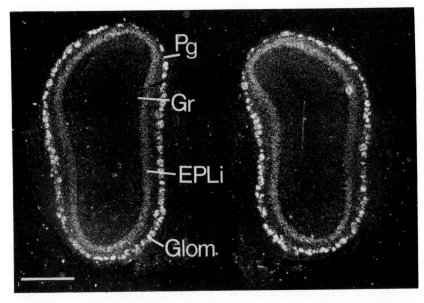

FIGURE 1. Dark-field photomicrograph of a frontal section through the olfactory bulbs of the rat, 24 h after intraventricular injection of 100 μl αBuTX (10^{-8} M). Perfusion with mixed aldehydes. (All subsequent figures were taken from material initially prepared in this fashion.) Label is heavily concentrated over the glomeruli (Glom) and inner portion of the external plexiform layer (EPLi). Granule (Gr) and periglomerular cell bodies (Pg) are fairly free of label. Exposure time, 1 week. Scale bar = 1 mm.

periglomerular neurons form a mixed population of cells containing the synthetic machinery to release either GABA or dopamine (Ribak, Vaughn, Saito, Barber, and Roberts, 1977; Halász, Ljungdahl, Hökfelt, Johansson, Goldstein, Park, and Biberfeld, 1977). Mitral/ tufted cells (and they will be considered as a single population from here on) therefore remain the only major population of olfactory bulb cells unaccounted for histochemically. If αBuTX binding is indicative of cholinergic transmission within the bulb, then the observed toxin patterns may be indicative of mitral-periglomerular and mitral-granule cell (excitatory) interactions that have been described both anatomically (Price and Powell, 1970*a*; Pinching and Powell, 1971*a, b*) and physiologically (Shepherd, 1972). Extrinsic cholinergic inputs could conceivably have been responsible for the binding pattern. However, there is evidence to suggest that the olfactory nerve is not cholinergic (Margolis, Roberts, Ferreiro, and Feldman, 1974). Centrifugal fibers which may be cholinergic (Broadwell and Jacobowitz, 1976) terminate on granule cells and within the periglomerular regions of the olfactory bulb (Price and Powell, 1970*b*), areas which do not bind toxin.

Having formulated the working hypothesis that mitral cells are cholinergic, we pursued the problem at the ultrastructural level, again using autoradiographic techniques. The interpretation of the data (which were principally gathered from observations on the glomerular layer) was made considerably easier by reference to the work of Pinching and Powell (1971*a, b*). In essence, these authors established that within the glomerulus, mitral-periglomerular synaptic contacts are always asymmetric (Gray type I) and formed by the *dendrites* of the two cell types. Synapses of the reverse polarity, between periglomerular and mitral cell dendrites, are always symmetrical (Gray type II). The presynaptic dendrites of the mitral cell were characterized by small groups of spherical vesicles associated with the synaptic specialization, while presynaptic periglomerular cell dendrites contained large numbers of predominantly flattened vesicles and abundant profiles of smooth endoplasmic reticulum. Both dendritic types contained ribosomes either free or in rosettes. Granule cell-mitral cell synaptic articulations are found only within the EPL and will therefore not be discussed here. A number of axon terminals have previously been found within the glomerular and periglomerular regions. These are derived from various sources but have never been seen in association with silver grains in electron microscopic autoradiographs.

Using the criteria for dendritic and synaptic identification described above, it became clear immediately that silver grains (and they were rarely in clusters) were almost never associated with olfactory nerve synapses and that the majority of grains were found in association with *dendritic* membranes, in some cases demonstrating synaptic specializations. In the majority of cases, vesicles were found in association with the labeled membranes, but as both pre- and postsynaptic profiles may contain such vesicles, a synaptic specialization was essential for the correct identification of synaptic polarity and thus of the nature of the postsynaptic cell to which toxin is presumed to bind (Hunt and Schmidt, 1978a).

Of 219 silver grains or silver grain clusters examined, 211 were associated with underlying dendritic membranes and were not found in relation to olfactory nerve or other axon terminals. Of the grains lying over dendritic membranes, 101 were found at the sites of apposition of mitral and periglomerular cell dendritic membranes. In a further 48 cases, asymmetric synaptic thickenings were seen, and presynaptic predominantly spherical vesicles were found within the mitral cell component of the synaptic complex. Silver grains were never found in association with unequivocally identified periglomerular-mitral cell synapses. Several labeled mitral-periglomerular synapses are illustrated in Figure 2. In one fortunate case, one limb of a reciprocal synapse between the mitral and periglomerular cell was labeled, while the periglomerular-mitral cell component was unlabeled (Figure 3).

FIGURE 2. Electron microscopic autoradiographs showing the association between silver grains and underlying mitral-periglomerular dendrodendritic synapses. Dm, mitral cell dendrite; Dp, periglomerular cell dendrite; s, spherical vesicles; ser, smooth endoplasmic reticulum.

A: Four dendritic profiles embedded in a dense matrix of olfactory nerve terminals (O). In the lower righthand corner an olfactory nerve synapse can be seen (arrows). Scale bar = 0.5 μm. This and subsequent EM photomicrographs taken from the glomerulus. B: The silver grain overlies an indistinctly resolved synaptic articulation (arrow), but the presence of large spherical vesicles would suggest that the mitral cell dendrite is establishing synaptic contact at this point. Scale bar = 0.5 μm. C: Silver grain, adjacent to an asymmetric synaptic thickening (arrow). Scale bar = 0.25 μm. D: Asymmetric synaptic contact with associated spherical presynaptic vesicles. Scale bar = 0.125 μm. E: A synaptic contact between a presumed mitral cell dendrite and periglomerular cell dendrite. Notice that while the vesicles are predominantly large and spherical, an occasional flattened profile can be seen. Scale bar = 0.125 μm.

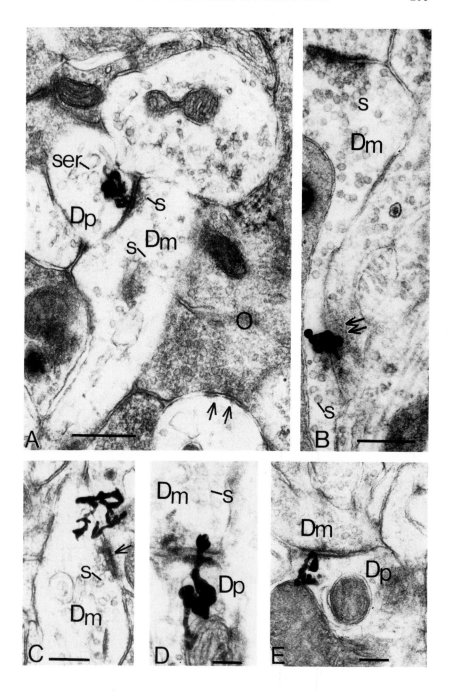

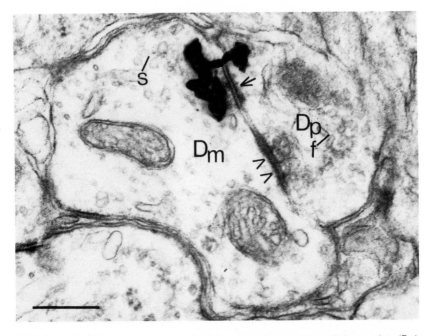

FIGURE 3. Reciprocal synapse between mitral (Dm) and periglomerular (Dp) dendrites within the glomerulus. Only the mitral-periglomerular (excitatory) synapse is labeled (arrows). The inhibitory limb of the dual synaptic arrangement (arrowheads), from the periglomerular dendrite (with flattened vesicles [f]) to the mitral cell (with spherical vesicles [s]), is unlabeled. Scale bar = 0.5 μm.

Esterase

The ultrastructural localization of toxin binding sites provided one piece of evidence in favor of the cholinergic hypothesis. We next examined the distribution of the enzyme acetylcholinesterase (AChE) within the olfactory bulb. AChE is found within both cholinergic and noncholinergic neurons (Butcher, Talbot, and Bilezikjian, 1975; Parent and Butcher, 1976), and perhaps the most forceful statement of the significance of the presence of the enzyme is that it is indicative of "sites at which cholinergic mechanisms may operate" (Silver, 1967). Nevertheless, if mitral cells are cholinergic, then these neurons should be able to synthesize the enzyme, and it should be possible to locate it histochemically. However, the distribution of AChE within the olfactory bulb is not at first sight compatible with this hypothesis. As has been previously demonstrated (Shute and Lewis, 1967; Sharma, 1968), AChE is found within the glomerular layer, inner portion of the EPL, and in a heavily staining submitral cell plexus of fibers (Figure 4A). The mitral cells appear to contain low levels of enzyme. We followed this problem up by looking for the synthesis sites of AChE using the technique of Butcher et al. (1975). In brief, rats are treated 6–20 hours before sacrifice with di-isopropyl fluorophosphate (DFP). This agent irreversibly inactivates brain AChE, which must be resynthesized within neuron cell bodies. Histochemical examination of the olfactory bulb after such treatment indicated that mitral cell bodies were indeed sites of AChE synthesis within the olfactory bulb (Figure 4B).

There are several other indications of the possible cholinergic nature of mitral cells. The distribution of the specific synthesizing enzyme for ACh, choline acetyltransferase (CAT), has been shown by microdissection to be predominantly within the glomerular and external plexiform layers (Godfrey, Ross, and Williams, 1977). ACh iontophoretically applied to mitral cells resulted, in the majority of cases, in the depression of the spontaneous firing rate, an effect antagonized by both nicotinic and muscarinic drugs (Salmoiraghi, Bloom, and Costa, 1964). The inhibition could have been the result of the direct excitation of granule cells, which are known to inhibit mitral cells by the release of the neurotransmitter GABA (McLennan, 1971; Nicoll, 1971).

One of the more useful aspects of the AChE histochemical technique is that surgical interruption of a cholinergic pathway produces a rapid fall in the terminal concentration of esterase (Shute and Lewis, 1961). For example, medial septal lesions which disrupt the septohippocampal pathway, perhaps the best-established cholinergic pathway within the central nervous system, result in the almost complete loss of AChE from within the hippocampal complex (Lewis, Shute, and Silver, 1967; Mellgren and Srebro, 1973). Applying the same paradigm to the presumptive mitral cell cholinergic pathway, ablation of the olfactory bulb should result in the loss of enzyme from the olfactory cortical areas within which mitral cell axons terminate. However, 5 and 7 days following olfactory bulb ablation we were unable to detect a decrease in staining intensity within layer I of the piriform cortex, within which mitral cell axons terminate (Price, 1973; Haberly and Price, 1977). Similarly, Harvey, Scholfield, Graham, and Aprison (1975) failed to find a significant drop in total pool size of acetylcholine within the piriform cortex.

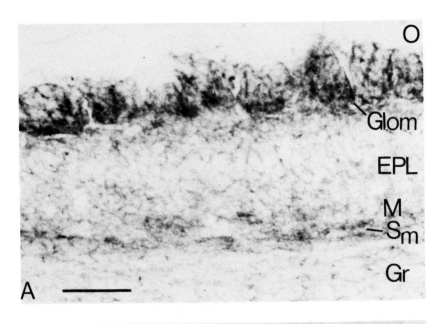

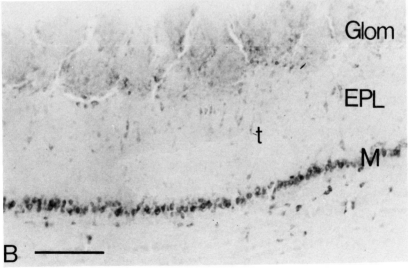

Other Candidates

While the evidence that ACh is released from mitral cell axons is rather weak, several lines of investigation have implicated either glutamate or aspartate as the neurotransmitter released from mitral cell axon terminals. Yamamoto and Matsui (1976) have demonstrated that exogenously applied [³H]glutamate can be released from in vitro slice preparations of guinea pig olfactory cortex following lateral olfactory tract (LOT) stimulation. Similarly, Bradford and Richards (1976) have reported the release of endogenous glutamate from a similar preparation. Section of the LOT results, after 8–9 days, in a fall in the total tissue concentrations of aspartate (31% reduction) and glutamate (20% reduction) within the olfactory cortex (Harvey et al., 1975), while Graham and Pong (1973) (quoted in the previous reference) found that glutamate levels were highest within the EPL, mitral, and granule cell layers of the olfactory bulb. Synaptosomes derived from olfactory bulb tissue also exhibited high-affinity uptake for aspartate and glutamate (Osborne, Duce, and Keen, 1976). In agreement with these observations, fresh tissue slices of olfactory cortex incubated in 10^{-6} M [³H]glutamate resulted in a pattern of labeling that was entirely consistent with the uptake and release of glutamate by mitral cell axons. Thus, the heaviest uptake of glutamate as judged from the autoradiographic reaction was over the most superficial cortical layer within which mitral cell axons terminate (unpublished observations).

CONCLUSION

It is possible, within the framework of our current knowledge (or rather, ignorance) concerning the neurotransmitter utilized by the mitral cell, to construct at least three models to accommodate the observations made.

(1) *Mitral cells are cholinergic.* The assumption that mitral cells are

---

FIGURE 4. Bright-field photomicrograph of a portion of the rat olfactory bulb stained for acetylcholinesterase. A: The normal case; B: 20 h after treatment with DFP. Note the heavy staining of mitral (M) and tufted (t) cell bodies in B but not A. As was the case with toxin binding sites, esterase is found predominantly within the inner portion of the EPL. Occasionally, positive cells are seen within the granule cell layer (Gr) and in the periglomerular region. Glom, glomeruli; O, olfactory nerve; Sm, submitral cell plexus of esterase-positive fibers. Frontal sections, 80 μm thick. Scale bar = 200 μm.

cholinergic agrees well with a number of observations, including the ones reported here, but is difficult to reconcile with the absence of any effect of the interruption of the LOT on esterase levels in the area of termination. We would then have to abandon the belief that, in the central nervous system, AChE is a presynaptic marker that regularly disappears from the target area when the presynaptic element of a cholinergic synapse degenerates. Persistence of esterase levels could for instance result from other cholinergic input into the area of termination of the LOT. Alternatively, as is known for the neuromuscular junction, a significant fraction of the esterase may be synthesized by the cholinoceptive cell and therefore survive removal of cholinergic terminals. As a further difficulty of the cholinergic model, the evidence in favor of glutamate as the mitral transmitter would remain unaccounted for.

(2) *Mitral cells are glutamate-ergic.* The hypothesis that mitral cells release glutamate as their transmitter readily explains the results of several experiments performed on the LOT, especially the pronounced uptake in the area of termination within the piriform cortex. It is also compatible with the capacity of the mitral cell to synthesize AChE, since the enzyme has been found in noncholinergic neurons. However, it would then become difficult to maintain the view that αBuTX binds exclusively to AChR in the brain, and a different function for αBuTX receptors—at least within the olfactory glomerulus—would have to be found.

(3) *Mitral cells release ACh and glutamate.* The proposition that mitral cells utilize both ACh and glutamate at the same time violates Dale's principle, according to which nerve cells release one and the same neurotransmitter from all of their branches. Several possibilities could be considered. In one version, glutamate and ACh are released together at all terminals, one serving as a transmitter and the other as a transmitter, auxiliary transmitter, or "modulator." The idea that ACh may serve such a role in noncholinergic synapses is of course not new. Abrahams, Koelle, and Smart proposed in 1957 that secretory fibers in the neurohypophysis might liberate ACh at their terminals to stimulate release of peptide hormones from the same terminals. The concept of a "cholinergic link" was later applied to synaptic transmission (Koelle, 1962) and especially to adrenergic synapses (Burn and Rand, 1965). In such a scheme, ACh would act at presynaptic sites, implying that toxin receptors likewise should be located in the presynaptic membrane; unfortunately, from our autoradiographic data it

is impossible to tell whether αBuTX binds to the pre- or post-synaptic membrane. One could argue that the term "neurotransmitter" should be reserved to messengers with postsynaptic activity (Burnstock, 1976), i.e., to glutamate in this particular model. A violation of Dale's principle would thereby be avoided. Another model might envisage a more radical departure from this principle by postulating that the mitral cell axon releases glutamate while the dendrites within the bulb utilize ACh. It should be pointed out that an extrasynaptic action of ACh must be considered as an explanation for the presence of silver grains over apparently nonsynaptic areas of mitral-periglomerular cell membrane apposition. Ramon-Moliner (1977) has recently postulated that GABA released from granule cells may act extrasynaptically upon mitral cells.

ACKNOWLEDGMENTS

This research was supported by a grant from the Council of Tobacco Research to J.S. S.H. was partly supported by National Institutes of Health grant NS-12078 to H. J. Karten, to whom we are greatly indebted for the use of facilities. We thank N. Brecha for reading the manuscript.

REFERENCES

Abrahams, V. C., G. B. Koelle, and P. Smart (1957). Histochemical demonstration of cholinesterases in the hypothalamus of the dog, *J. Physiol. (London)* **139**:137–144.

Barnard, E. A., J. O. Dolly, C. W. Porter, and E. X. Albuquerque (1975). The acetylcholine receptor and the ionic conductance modulation system of skeletal muscle, *Exp. Neurol.* **48**:1–28.

Bartfai, T., P. Berg, M. Schultzberg, and E. Heilbronn (1976). Isolation of a synaptic membrane fraction enriched in cholinergic receptors by controlled phospholipase A_2 hydrolysis of synaptic membranes, *Biochim. Biophys. Acta* **426**:186–197.

Bradford, H. F. and C. D. Richards (1976). Specific release of endogenous glutamate from piriform cortex stimulated *in vitro, Brain Res.* **105**: 168–172.

Broadwell, R. D. and D. M. Jacobowitz (1976). Olfactory relationships of the telencephalon and diencephalon in the rabbit. III. The ipsilateral centrifugal fibers to the olfactory bulbar and retrobulbar formations, *J. Comp. Neurol.* **170**:321–346.

Brown, D. A. and L. Fumagalli (1977). Dissociation of α-bungarotoxin binding and receptor block in the rat superior cervical ganglion, *Brain Res.* **129**:165–168.

Burn, J. H. and M. J. Rand (1965). Acetylcholine in adrenergic transmission, *Annu. Rev. Pharmacol.* **5:**163–182.

Burnstock, G. (1976). Do some nerve cells release more than one transmitter? *Neuroscience* **1:**239–248.

Butcher, L. L., K. Talbot, and L. Bilezikjian (1975). Acetylcholinesterase neurons in dopamine-containing regions of the brain, *J. Neural Transm.* **37:**127–153.

Devreotes, P. N. and D. M. Fambrough (1975). Turnover of acetylcholine receptors in skeletal muscle, *Cold Spring Harbor Symp. Quant. Biol.* **40:** 237–251.

Duggan, A., J. G. Hall, and C. Y. Lee (1976). Alpha-bungarotoxin, cobra neurotoxin, and excitation of Renshaw cells by acetylcholine, *Brain Res.* **107:**166–170.

Eterović, V. A. and E. L. Bennett (1974). Nicotinic cholinergic receptor in brain detected by binding of α-[^{3}H]-bungarotoxin, *Biochim. Biophys. Acta* **362:**346–355.

Fertuck, H. C. and M. M. Salpeter (1974). Localization of acetylcholine receptor by ^{125}I-labeled α-bungarotoxin binding at mouse motor endplates, *Proc. Natl. Acad. Sci. U.S.A.* **71:**1376–1378.

Fewtrell, C. M. S. (1976). The labelling and isolation of neuro-receptors, *Neuroscience* **1:**249–273.

Freeman, J. A. (1977). Possible regulatory function of acetylcholine receptor in maintenance of retinotectal synapses, *Nature (London)* **269:**218–222.

Godfrey, D. A., C. D. Ross, and A. D. Williams (1977). Choline acetyltransferase and acetylcholinesterase in the olfactory system of the rat, *Abstr. Annu. Meet. Soc. Neurosci.,* 7th, Anaheim, p. 78.

Haberly, L. B. and J. L. Price (1977). The axonal projection patterns of the mitral and tufted cells of the olfactory bulb in the rat, *Brain Res.* **129:** 152–157.

Halász, N., A. Ljungdahl, T. Hökfelt, O. Johansson, M. Goldstein, D. Park, and P. Biberfeld (1977). Transmitter histochemistry of the rat olfactory bulb. I. Immunohistochemical localization of monoamine synthesizing enzymes. Support for intrabulbar, periglomerular dopamine neurons, *Brain Res.* **126:**455–474.

Harvey, J. A., C. N. Scholfield, L. T. Graham, Jr., and M. H. Aprison (1975). Putative transmitters in denervated olfactory cortex, *J. Neurochem.* **24:** 445–449.

Hunt, S. P. and J. Schmidt (1978*a*). The electron microscopic autoradiographic localization of α-bungarotoxin binding sites within the central nervous system of the rat, *Brain Res.* **142:**152–159.

Hunt, S. P. and J. Schmidt (1978*b*). Some observations on the binding patterns of α-bungarotoxin in the central nervous system of the rat, *Brain Res.,* in press.

Kása, P. and A. Silver (1969). The correlation between choline acetyltransferase and acetylcholinesterase activity in different areas of the cerebellum of rat and guinea pig, *J. Neurochem.* **16:**389–396.

Koelle, G. B. (1962). A new general concept of the neurohumoral function of acetylcholine and acetylcholinesterase, *J. Pharm. Pharmacol.* **14:**65–90.

Koelle, G. B. (1970). Neurohumoral transmission and the autonomic nervous system, pp. 402–441 in *The Pharmacological Basis of Therapeutics,* Goodman, L. S. and A. Gilman, eds. Macmillan Publishing Co., New York.

Lentz, T. L. and J. Chester (1977). Localization of acetylcholine receptors in central synapses, *J. Cell Biol.* **75**:258–267.

Lentz, T. L., J. E. Mazurkiewicz, and J. Rosenthal (1977). Cytochemical localization of acetylcholine receptors at the neuromuscular junction by means of horseradish peroxidase-labeled α-bungarotoxin, *Brain Res.* **132**: 423–442.

Lewis, P. R., C. C. D. Shute, and A. Silver (1967). Confirmation from choline acetylase analyses of a massive cholinergic innervation to the rat hippocampus, *J. Physiol. (London)* **191**:215–224.

Lowy, J., J. MacGregor, J. Rosenstone, and J. Schmidt (1976). Solubilization of an α-bungarotoxin-binding component from rat brain, *Biochemistry* **15**: 1522–1527.

Margolis, F. L., N. Roberts, D. Ferreiro, and J. Feldman (1974). Denervation in the primary olfactory pathway of mice: biochemical and morphological effects, *Brain Res.* **81**:469–483.

McLennan, H. (1971). The pharmacology of inhibition of mitral cells in the olfactory bulb, *Brain Res.* **29**:177–184.

Mellgren, S. I. and B. Srebro (1973). Changes in acetylcholinesterase and distribution of degenerating fibres in the hippocampal region after septal lesions in the rat, *Brain Res.* **52**:19–36.

Miledi, R. and A. C. Szczepaniak (1975). Effect of Dendroaspis neurotoxins on synaptic transmission in the spinal cord of the frog, *Proc. R. Soc. London B Biol. Sci.* **190**:267–274.

Nicoll, R. A. (1971). Pharmacological evidence for GABA as the transmitter in granule cell inhibition in the olfactory bulb, *Brain Res.* **35**:135–149.

Osborne, R. H., I. R. Duce, and P. Keen (1976). Amino acids in "light" and "heavy" synaptosome fractions from rat olfactory lobes and their release by electrical stimulation, *J. Neurochem.* **27**:1483–1488.

Parent, A. and L. L. Butcher (1976). Organization and morphologies of acetylcholinesterase-containing neurons in the thalamus and hypothalamus of the rat, *J. Comp. Neurol.* **170**:205–226.

Patrick, J. and W. B. Stallcup (1977). Immunological distinction between acetylcholine receptor and the α-bungarotoxin-binding component on sympathetic neurons, *Proc. Natl. Acad. Sci. U.S.A.* **74**:4689–4692.

Pinching, A. J. and T. P. S. Powell (1971a). The neuropil of the periglomerular region of the olfactory bulb, *J. Cell Sci.* **9**:379–409.

Pinching, A. J. and T. P. S. Powell (1971b). The neuropil of the glomeruli of the olfactory bulb, *J. Cell Sci.* **9**:347–377.

Polz-Tejera, G., J. Schmidt, and H. J. Karten (1975). Autoradiographic localization of α-bungarotoxin-binding sites in the central nervous system, *Nature (London)* **258**:349–351.

Price, J. L. (1973). An autoradiographic study of complementary laminar patterns of termination of afferent fibers to the olfactory cortex, *J. Comp. Neurol.* **150**:87–108.

Price, J. L. and T. P. S. Powell (1970a). The synaptology of the granule cells of the olfactory bulb, *J. Cell Sci.* **7**:125–155.

cortical regions (Collins, Kennedy, Sokoloff, and Plum, 1976). These and other studies have given confidence that the autoradiographic density patterns revealed by the DG method are a valid reflection of metabolic activity related to functional properties of neurons.

The olfactory bulb was one of the first brain regions to which the DG method was applied (Sharp, Kauer, and Shepherd, 1975; Shepherd, 1976). The rationale was that the olfactory bulb is largely composed of neuropil, with a high density of synaptic connections between axonal and dendritic terminals. Synaptic transmission was known to have relatively high energy demands (cf. Dolivo and Roullier, 1969), and the olfactory bulb therefore appeared to be well suited to the method.

The particular problem we addressed was that of the spatial distribution of activity in the olfactory bulb during stimulation with odors. Spatial processing mechanisms are important in most sensory systems; indeed, much of our understanding of sensory mechanisms is related to the orderly transmission of spatial aspects of sensory stimuli. However, in the olfactory system there is no obvious spatial component of the stimulus. Electrophysiological experiments have provided suggestive evidence of spatial gradients of activity (Adrian, 1953; Mozell, 1964), receptive fields (Kauer and Moulton, 1974), and selective degeneration (Doving and Pinching, 1973) during odor stimulation, but the results, particularly in the mammal, have been difficult to interpret. Our aim therefore was to see whether the DG method would provide evidence regarding the spatial distribution of activity associated with odor stimulation, evidence that had eluded more traditional methods and that might be used to guide future functional analysis.

Our initial experiments were carried out following the procedures of Kennedy et al. (1975). Rats were injected with a single intravenous dose of [2-^{14}C]deoxyglucose in a tail vein and placed in a glass bottle through which control air or air containing the test odor circulated. After 45 minutes the animals were killed, and the bulbs were quickly removed, frozen, and sectioned in a cryostat at $-10°$ C. The sections were then dried and exposed to X-ray film for 5 days to give the autoradiographs. The histological sections were stained, and enlarged prints of the sections and the autoradiographs were made for detailed laminar localization and densitometric measurements of the density patterns.

In control rats exposed to room air, the basic density patterns consist of three concentric bands, as shown in Figure 1 (Sharp, Kauer, and Shepherd, 1975, 1977). There is an outer light band associated with the olfactory nerve layer; an intermediate dark band associated with the

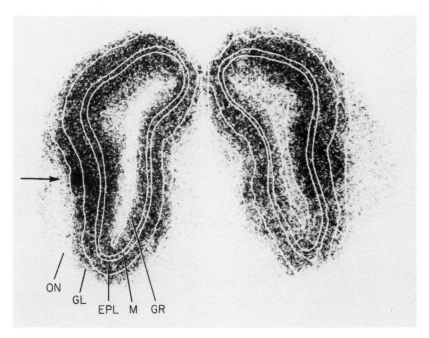

FIGURE 1. Autoradiograph of frontal sections of olfactory bulbs of a rat exposed only to room air. The outlines of the histological layers of the olfactory bulb, as determined from the subsequently stained sections, are shown superimposed on the autoradiographs. ON, olfactory nerve layer; GL, glomerular layer; EPL, external plexiform layer; M, mitral cell body layer; GR, granule layer. Arrow points to site of small, dense focus situated over the glomerular layer. Scale bar is 500 μm.

external plexiform layer and layers of olfactory glomeruli, mitral cell bodies, and granule cells; and an internal light band related to the periventricular core. In addition, very small, dense foci are characteristically found scattered through the bulb, as illustrated in Figure 1. Careful correlation with the histological laminae shows that these foci lie precisely over small groups of olfactory glomeruli.

In rats exposed to the odor of amyl acetate, larger regions of increased density are characteristically found in the olfactory bulb, as illustrated in Figure 2 (Sharp et al., 1977). These regions tend to lie in the lateral and medial parts of the bulb. Densitometric measurements correlated with the histological layers show that there is a peak of density in the glomerular layer, with extension into the neighboring olfactory nerve and external plexiform layers. The increased density sometimes appears

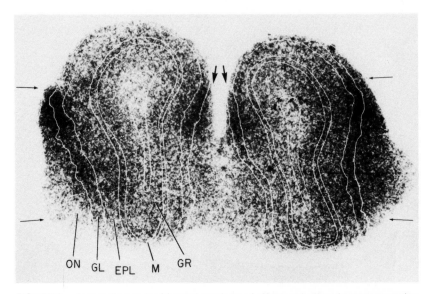

FIGURE 2. Autoradiograph of frontal sections of olfactory bulbs of a rat exposed to strong odor of amyl acetate. Superimposition of outline of the histological layers, as in previous figure. Small arrows indicate extent of lateral active regions; large arrows indicate medial active regions. Scale bar is 500 μm.

to extend to the deeper mitral and granule cell layers, as in Figure 2. We are currently testing the effects of different odors (Stewart, Kauer, and Shepherd, 1977), and preliminary results suggest that these produce autoradiographic patterns which differ in their topographical distribution. Comparable results have been reported in the tree shrew (Skeen, 1977).

The foregoing observations apply to the olfactory bulb. In the olfactory cortex, the autoradiographic densities also have a characteristic laminar pattern (Sharp et al., 1977). There is an outer light band associated with the nerve fibers in the lateral olfactory tract; a broad, dense band which has its peak in the molecular layer and extends into the deeper-lying pyramidal and polymorphic cell layers; and an inner light zone associated with the white matter. We have not observed any topographical differences in density within the olfactory cortex in relation to odor stimulation.

Several conclusions are suggested by these initial studies. In the olfactory bulb, the peak of activity in the glomerular layer is correlated with the high density of axonal and dendritic terminals in that layer (cf.

Sharp, 1976). In this layer the axons of the receptor cells make synaptic connections onto the dendrites of bulbar neurons—mitral, tufted, and periglomerular cells (Pinching and Powell, 1971). There is an analogy in this respect with lamina IV in the cerebral cortex, where thalamic afferents make synaptic connection onto cortical cells; the DG laminar densities show a similar peak in this layer in visual cortex (Kennedy, Des Rosiers, Sakurada, Shinohara, Reivich, Jehle, and Sokoloff, 1976). In the olfactory glomeruli there are also synaptic connections between the dendritic terminals (Pinching and Powell, 1971; White, 1973); their contribution to glucose uptake remains to be assessed.

The very small foci of activity in control animals breathing room air have several interesting implications. Since there was no obvious odor in the ambient air, these foci appear to represent activity at very low levels of olfactory stimulation. These foci were completely unexpected; there was little evidence from previous studies for this type of activity. The finding of these foci in turn suggests that the DG method can be an effective means for detecting relatively low levels of nervous activity in the brain. In fact, it appears that these foci are among the most sensitive indicators yet known of natural physiological activity in awake, behaving animals.

The larger regions of activity induced by odor stimulation indicate that the mammalian olfactory bulb is not spatially homogeneous with respect to the processing of odor information. The fact that the patterns appear to be different for different odors suggests that they may be involved in the mechanisms for discrimination between odors. It should be emphasized in this regard that the patterns do not have the rigid geometrical arrangement that characterizes the topographical organization of other sensory systems such as the visual or the somatosensory. This, of course, may simply reflect the fact that olfactory processing does not operate within the kind of external spatial coordinates that are imposed on other systems. But the variability of the patterns in the olfactory bulb may also reflect the sensitivity of the DG method, as mentioned above, and the effects of behavioral variables (such as alterations in intranasal odor currents) that we have not yet adequately controlled.

It is tempting to speculate that the regional patterns of glucose uptake during odor stimulation are due to inputs to the glomeruli from specific functional groups of receptor cells in the nose. We have as yet no evidence for such organization at the receptor level using the DG method. However, studies of the olfactory nerve projection to the olfactory bulb using degeneration (Land, 1973) and amino acid transport

(Land and Shepherd, 1974) have demonstrated topographical relations between parts of the nasal cavity and parts of the olfactory bulb, and, within these regions, projections of specific olfactory nerve bundles to specific olfactory glomeruli. In addition, electrophysiological studies in the salamander using punctate odor stimulation of the receptor sheet have demonstrated restricted receptive fields for mitral cells in the olfactory bulb (Kauer and Moulton, 1974).

From these results it appears that the DG method has provided evidence regarding the possible role of spatial activity patterns for information processing in the olfactory system. The advantages of the DG method are well illustrated by its application to this particular system: its sensitivity; its use in awake, normally behaving animals; the serial maps of simultaneous activity throughout the region under study; and the evidence for multiple sites of activity that were not previously suspected from other studies and that would be difficult to demonstrate with electrophysiological recordings. Given this initial evidence, considerable work remains to be done before we understand the mechanisms and significance of these activity patterns and can relate them to the results of analyses of olfactory bulb organization by other methods.

REFERENCES

Adrian, E. D. (1953). Sensory messages and sensation: the response of the olfactory organ to different smells, *Acta Physiol. Scand.* **29:**5–14.

Collins, R. C., C. Kennedy, L. Sokoloff, and F. Plum (1976). Metabolic anatomy of focal motor seizures, *Arch. Neurol.* **33:**536–542.

Dolivo, M. and M. Roullier (1969). Changes in ultrastructure and synaptic transmission in the sympathetic ganglion during various metabolic conditions, *Brain Res.* **31:**111–123.

Doving, K. B. and A. J. Pinching (1973). Selective degeneration of neurones in the olfactory bulb following prolonged odour exposure, *Brain Res.* **52:**115–129.

Hubel, D. H., T. N. Wiesel, and M. P. Stryker (1977). Orientation columns in macaque monkey visual cortex demonstrated by the 2-deoxyglucose autoradiographic technique, *Nature (London)* **269:**328–330.

Kauer, J. S. and D. G. Moulton (1974). Responses of olfactory bulb neurones to odour stimulation of small nasal areas in the salamander, *J. Physiol. (London)* **243:**717–737.

Kennedy, C., M. H. Des Rosiers, J. W. Jehle, M. Reivich, F. R. Sharp, and L. Sokoloff (1975). Mapping of functional neural pathways by autoradiographic survey of local metabolic rate with [^{14}C]deoxyglucose, *Science* **187:**850–853.

Kennedy, C., M. H. Des Rosiers, O. Sakurada, M. Shinohara, M. Reivich, J. W. Jehle, and L. Sokoloff (1976). Metabolic mapping of the primary visual system of the monkey by means of the autoradiographic [^{14}C]deoxyglucose technique, *Proc. Natl. Acad. Sci. U.S.A.* **73:**4230–4234.

Land, L. J. (1973). Localized projection of olfactory nerves to rabbit olfactory bulb, *Brain Res.* **63:**153–166.

Land, L. J. and G. M. Shepherd (1974). Autoradiographic analysis of olfactory receptor projections in the rabbit, *Brain Res.* **70:**506–510.

Mozell, M. M. (1964). Olfactory discrimination: electrophysiological spatio-temporal basis, *Science* **143:**1336–1337.

Pinching, A. J. and T. P. Powell (1971). The neuropil of the glomeruli of the olfactory bulb, *J. Cell Sci.* **9:**347–377.

Sharp, F. R. (1976). Relative cerebral metabolic rates of neuron perikarya and neuropil determined with 2-deoxyglucose in resting and swimming rat, *Brain Res.* **110:**127–140.

Sharp, F. R., J. S. Kauer, and G. M. Shepherd (1975). Local sites of activity-related glucose metabolism in rat olfactory bulb during olfactory stimulation, *Brain Res.* **98:**596–600.

Sharp, F. R., J. S. Kauer, and G. M. Shepherd (1977). Laminar analysis of 2-deoxyglucose uptake in olfactory bulb and olfactory cortex of rabbit and rat, *J. Neurophysiol.* **40:**800–813.

Shepherd, G. M. (1976). Olfactory stimulation, in Neuroanatomical Functional Mapping as Determined by the Radioactive 2-Deoxy-D-glucose Method, *Neurosci. Res. Program Bull.* **14:**478–484.

Skeen, L. C. (1977). Odor-induced patterns of deoxyglucose consumption in the olfactory bulb of the tree shrew, *Tupaia glis, Brain Res.* **124:**147–153.

Sokoloff, L. (1975). Influence of functional activity on local cerebral glucose utilization, pp. 385–388 in *Brain Work: The Coupling of Function, Metabolism and Blood Flow in the Brain,* Ingvar, D. H. and N. A. Lassen, eds. Academic Press, New York.

Stewart, W. B., J. S. Kauer, and G. M. Shepherd (1977). 2-Deoxyglucose uptake patterns in rat olfactory bulb under different odor conditions, *Abstr. Annu. Meet. Soc. Neurosci.*, 7th, Anaheim, p. 84.

White, E. L. (1973). Synaptic organization of the mammalian olfactory glomerulus: new findings including an intraspecific variation, *Brain Res.* **60:**299–313.

HORMONE BINDING, OLFACTORY PATHWAYS, AND HORMONE-CONTROLLED BEHAVIORS

Donald W. Pfaff

Rockefeller University, New York, New York

INTRODUCTION

Olfactory input has been shown to influence a variety of natural behavior patterns which are also controlled by steroid hormones. Input from the vomeronasal organ has, for many decades, been suspected to be important for such behavior patterns, although definitive evidence for this notion has begun to accumulate only recently. For these reasons, it is interesting that nerve cells in the basal telencephalon accumulate steroid sex hormones (first section below). In turn, some of these neurons which receive an olfactory or vomeronasal input (from the olfactory bulb proper or from the accessory olfactory bulb) project to the preoptic area or hypothalamus. There they can influence the modulation of masculine sex behavior, female-typical behavior, and autonomic function (second section below). Finally, neurons in the preoptic area, bed nucleus of the stria terminalis, and basomedial hypothalamus, some of which bind estrogenic or androgenic hormones, project back to the telencephalon (third section below). These anatomical connections afford the possibility of a hormone-sensitive centrifugal control on olfactory or vomeronasal input.

HORMONE-CONCENTRATING CELLS IN OLFACTORY PATHWAYS

A considerable amount of neuroanatomical work, summarized in this symposium by Broadwell (1978), has shown that the accessory olfactory bulb, which receives its input from the vomeronasal organ, projects to the medial nucleus of the amygdala and the most medial portion of the cortical nucleus of the amygdala. Since this type of chemosensory input has long been suspected to be important for the control of instinctive behavior, it is very interesting that this portion of the amygdala includes nerve cells which bind estrogenic or androgenic hormones. For two examples of this, we can turn to the demonstration of estrogen-binding nerve cells in the medial amygdaloid nucleus of the female rat (Pfaff and Keiner, 1973) (note also that some estrogen-binding cells can be found in the prepiriform cortex, which receives input from the olfactory bulb proper), and the demonstration of estrogen-binding cells in the medial amygdala of the rhesus monkey (Pfaff, Gerlach, McEwen, Ferin, Carmel, and Zimmerman, 1976). Steroid sex hormone binding in the amygdala seems to be a strong phenomenon with wide phylogenetic generality. We have investigated the accumulation of radioactive estrogen and androgen by nerve cells in a wide variety of vertebrate species and have arrived at three generalizations about sex hormone binding in the brain which appear to be true of all vertebrates (Morrell, Kelley, and Pfaff, 1975; Pfaff, 1976). First, in all species examined we were able to find in specific locations nerve cells which bound estradiol or testosterone. Second, there was a core of phylogenetically ancient structures which always included estrogen- or androgen-concentrating neurons. These phylogenetically ancient structures included limbic cell groups such as the medial amygdala or its homologous structure (in lower vertebrates), and also included the medial preoptic area and cell groups in the basomedial (tuberal) hypothalamus. Third, in a large majority of the cases for all of the animals studied, there was evidence that sex hormone-concentrating cell groups are involved in the control of hormone-dependent pituitary or behavioral function.

RELATIONS TO HORMONE-SENSITIVE
BEHAVIOR PATTERNS

Neuroanatomical projections from the amygdala, which receives olfactory and vomeronasal input, to the preoptic area, bed nucleus of the stria terminalis, and medial hypothalamus are crucial to understanding how hormone-influenced input could influence reproductive behavior patterns. In the rat, no convincing evidence has been found, using silver staining after experimental lesions, of a ventral pathway from the amygdala to the hypothalamus (Leonard and Scott, 1971). Therefore, the predominant influences from the amygdala to the preoptic area and hypothalamus must be carried via the stria terminalis. Lesions in the posterior part of the cortical nucleus of the amygdala, which receives a projection from the accessory olfactory bulb, led to degeneration in the supracommissural stria terminalis, projecting to the ventromedial region of the hypothalamus (Leonard and Scott, 1971). Neurons throughout a more widespread area in the amygdala give rise to the postcommissural stria terminalis, which projects to the bed nucleus of the stria terminalis and the preoptic area (Leonard and Scott, 1971). Thus, we can imagine a chain of neurons including either the corticomedial amygdala and the preoptic area or the cortico-medial amygdala and the ventromedial nucleus of the hypothalamus. In either case, based on the autoradiographic data mentioned above, it is clear that a group of cells which bind estrogenic or androgenic hormones is projecting to another group of cells which also bind these hormones. An implication of this is that a given steroid sex hormone is probably influencing nerve cells at more than one site in a functional pathway.

The exact importance of the stria terminalis for reproductive behavior and neuroendocrine events is not yet entirely clear. For instance, in female rats, complete bilateral transection of the stria terminalis does not prevent estrus cycling from continuing in a regular fashion. Female reproductive behavior in rats is neither abolished nor facilitated in any simple fashion by amygdaloid lesions, although there are some interactions between the amygdala and the septum in a joint contribution to female reproductive function. On the other hand, the work of Sachs and his collaborators has shown that lesions of the amygdala or of the bed nucleus of the stria terminalis can decrease the performance of male reproductive behavior in rats.

The preoptic and medial hypothalamic targets of the stria terminalis have clear relationships to reproductive behavior and autonomic function. Neurons in the medial preoptic area are important for the execution of male reproductive behavior and for the facilitation of parasympathetic function. Neurons and axons further posterior among preoptic and hypothalamic cell groups, in and around the ventromedial nucleus of the hypothalamus and in the posterior hypothalamus, are important for the execution of female reproductive behavior and sympathetic autonomic function (reviewed by Pfaff and Modianos, 1977). The neural circuitry for the control of lordosis behavior in the female rodent, modulated by neurons in and around the ventromedial nucleus of the hypothalamus, is being worked out (Pfaff and

NOTES

J. D. GRABOW, M.D.